Célia Balate Langa

# Influence of environmental conditions on aviation accidents involving birds

Célia Balate Langa

# Influence of environmental conditions on aviation accidents involving birds

The case of Maputo International Airport, between 2016 and 2021

ScienciaScripts

**Imprint**

Any brand names and product names mentioned in this book are subject to trademark, brand or patent protection and are trademarks or registered trademarks of their respective holders. The use of brand names, product names, common names, trade names, product descriptions etc. even without a particular marking in this work is in no way to be construed to mean that such names may be regarded as unrestricted in respect of trademark and brand protection legislation and could thus be used by anyone.

Cover image: www.ingimage.com

This book is a translation from the original published under ISBN 978-620-6-76004-7.

Publisher:
Sciencia Scripts
is a trademark of
Dodo Books Indian Ocean Ltd. and OmniScriptum S.R.L publishing group

120 High Road, East Finchley, London, N2 9ED, United Kingdom
Str. Armeneasca 28/1, office 1, Chisinau MD-2012, Republic of Moldova, Europe
Printed at: see last page
**ISBN: 978-620-7-87267-1**

# Contents

# <u>Dedication</u>

I dedicate this work to my daughters Patrícia Isabel and Daniela Larissa, the reason for my life.

"Knowing how to listen is the source of life"
(Author: José Simão Balate, my late father)

**Thanks s**

First of all to God, for the gift of life, for his help and protection, for his strength and constant presence, and for guiding me to the conclusion of yet another precious stage in my life.

To my dear professor and advisor Prof. Dr. José Julião da Silva, who dedicated his time and shared his experience so that my training would also be a life learning experience, my affection and my thanks. His critical and constructive eye helped me overcome the challenges of this dissertation, and I will be eternally grateful.

My thanks go to the FCTA faculty, and in particular to the lecturers who taught the modules of the master's degree in Environmental Risk Management, for their encouragement and attention during the course.

I would like to thank my former classmate, the first class of the Bachelor's and Licentiate's Degree in Geography Teaching (2006), from the post-work period at the Pedagogical University of Maputo, Ms. Rosalina Inácio Fumo Langa, for her encouragement and strength to return to academia, for the attention she gave when I applied and for the materials she shared.

To my husband Gonçalves Langa, who, with a great deal of trust, dedication, strength and love, made this dream come true. And even more so, for understanding the importance of this achievement and accepting my absence when necessary. Thank you very much.

To my children, Patrícia, Daniela and Beny, who, despite being small, accepted and understood the need for my absence very naturally.

I would like to thank my parents, José Simão Balate (in memory) and Maria Isabel, for the foundations they gave me to become the person I am today.

4

To all my fellow students (Manos), as we affectionately called each other, with whom I lived during these years of great learning, for their companionship and spirit of mutual help in carrying out the work.

To Engineer Manuel Vieira, responsible for Environmental Management at Aeroportos de Moçambique, E.P. for his guidance and supply of material for the research.

To Mr. José Candrinho (Director of Airport Operations) and Mr. Tomo Langa (Security Manager), employees of Maputo International Airport, for their availability and help in collecting and interpreting the data collected in the field.

To Dr. Neusia Machava, director of the Directorate of Economic Regulation at the Civil Aviation Authority of Mozambique, for her support and encouragement to take on the challenge of studying this subject and guidance in structuring the proposal for a regulation establishing the Control of Fauna in the vicinity of Airports and Aerodromes.

To all those who have been and are close to me in some way, making this life more and more worthwhile, thank you very much.

# <u>Summary</u>

Attention to the safety of air travel requires a focus on various adversities related to this event, such as aircraft collisions with fauna, especially birds. Of the various causes of this phenomenon, the problem related to the environment stands out, both around the airport and in the vicinity of aerodromes, as a potential factor in attracting these species to the locality, aggravating the problem of aviation safety. However, the relationship between land cover patterns and the risk of collision between aircraft and birds is still a gap in the literature, and needs to be investigated in order to promote public policies. The collision of birds with aircraft is technically called *Bird Strike*. This study, entitled Influence of environmental conditions on the occurrence of aviation accidents associated with birds: the case of Maputo International Airport between 2016 and 2021, set out to analyze the socio-environmental factors that influence the occurrence of collisions between aircraft and birds at Maputo International Airport between 2016 and 2021. In order to carry out this research, research methods and techniques were adopted. Among the methods used in this research are the statistical method, cartography, bibliographic research with the aim of identifying the main factors that contribute to the occurrence of collisions between birds and aircraft and documentary analysis, The research techniques used to collect data were also described, such as surveys, interviews, direct and indirect observation, as well as statistical methods. Finally, the methodology for analyzing avian risks is presented, based on two assessment models (Consultation of Collision Reports Methodology I and Methodology II proposals for problem species). According to the results of the research, there is no doubt that collision reporting is a way of increasing aviation safety at airports and in areas close to the airport. It is a fact that when something out of the ordinary (accident/incident) happens during a flight, everything needs to be reported as a way of creating an internal safety culture, helping to prevent accidents and improve internal processes. In relation to the risk assessment with the application of assessment methodology I and II, the results obtained allowed us to conclude that the rates of collision between birds and aircraft tend to increase even with the decrease in airport operations within Maputo International Airport, which shows that there is an increase in the frequency of bird visits within the ASA. Finally, it was possible to see that in terms of the level of risk looking at the problem species, Maputo International Airport presents a very high risk, and the problem species pointed out in this research are related to the focus of attraction existing in the ASA as well as inside the airport, this dissertation has annexes and appendices related to the instruments applied in the study area.

Keywords: Avian Risk, Airports, Collisions, Environmental Conditions, Aircraft.

<u>**0.Introduction**</u>

The air transport industry plays an important role in the world economy, with air traffic having increased significantly in recent years. This increase is expected to continue in the future, both in the medium and short term (ICAO 2016). Mozambique is no exception to this rapid increase in air traffic, with a total of 107,091 operations between 2016 and 2021, including embarkations and disembarkations, making it necessary to update the existing infrastructure to accommodate it. It is therefore essential that aviation safety planning keeps pace with this progress (ICAO 2016).

It should be noted that in order to respond to the rapid growth in air traffic, countries are obliged to make improvements to their airport facilities and to accommodate this increase in air traffic Mozambique has carried out projects to build new airports in the provinces of Nampula and Gaza (Nacala International Airport and Filipe Jacinto Nyusi, respectively) and to remodel Maputo, Pemba and Vilanculos airports.

According to ICAO (2016), aviation safety planning is essential in the process of building, upgrading and modernizing airports in order to avoid the many accidents caused by birds at airports, known as *bird strikes*[1] .

Figures provided by LAM (Linhas Aéreas de Moçambique, SA) show that it carried a total of 48,175 passengers in May 2018, compared to 36,646 in the same period in 2021, an increase of 31%.

The four routes with the highest demand between 2016 and 2021 were Nampula - Maputo with 8,442 passengers, Beira - Maputo 8,684, Maputo -

---

[1] ***Bird strikes are*** strictly defined as a collision between a bird and an aircraft that is in flight or in the take-off or landing phase. The term is often expanded to cover other wildlife attacks - with bats or land animals.

Pemba 5,804, Maputo - Tete with 5,944 passengers. It should also be noted that in the same month the national airline recorded an operational punctuality rate of 74% for a total of 871 departures, an increase of 5 percentage points compared to the same period in 2021, when punctuality was 69% for a total of 588 departures.

However, when compared to April 2022, there was a decrease of 4 percentage points, with a punctuality rate of 78% in this period. Among the stopovers analyzed in this period, the following stand out: Chimoio with 94%, Tete with 87%, Nacala with 86% and Vilankulo and Harare with 85% respectively, in terms of regularity[2] , punctuality[3] and efficiency operational[4] .

The existence of birds in areas close to airports has been seen as a constant risk factor for air operations due to their impact on aircraft when they are landing or taking off. *Birds* colliding with aircraft is commonly referred to as *Bird Strikes* (SAIUNDO, 2019:05).

This concern about collisions between aircraft and birds stems from the rapid                            growth                            in growth in air traffic, urban growth, disorderly occupation of the land around the airport, the construction and expansion of new airports that end up "invading" the habitat of animals, especially birds that have a dense population, leading to a probability and severity of collision, mainly because most records occur in the approach, take-off and landing procedures, that is, close to the ground,

---

[2] ***Regularity index:*** expresses the percentage of the total planned flight legs that were actually carried out, according to parameters defined by the Civil Aviation Authority.

[3] ***Punctuality indices:*** expresses the percentage of flight legs that departed on time out of those performed, according to parameters defined by the Civil Aviation Authority. In general terms, flights that:
- Domestic flights: engine start no later than 10 (ten) minutes before or 15 (fifteen) minutes after the scheduled time and/or engine stop no later than 15 (fifteen) minutes after the scheduled time.
*- International flights:* engine start within 30 (thirty) minutes before or after the scheduled time and/or engine stop within 30 (thirty) minutes after the scheduled time.

[4] **Operational efficiency index:** equivalent to the product of the two previous indices, corresponding to the combined action of regularity and punctuality.

considered the critical phases of a flight. (MENDONÇA, 2009 cited by FERNANDES, 2017:13).

According to Kalafatas (2010), aviation accidents involving collisions with wildlife represent the second largest cause of fatalities in civil aviation. The first bird strike accident resulting in death occurred in the United States of America in 1912, with American Calbraith Rogers, on a flight over Long Beach (LEWIS, 1995). Since that time, collisions with birds have caused the loss of 52 civil aircraft and approximately 190 lives worldwide (ALLAN, 2000). Some cases of avian danger are widely cited in aviation history. For example, US Airways flight 1549, whose Airbus 320 collided with Canadian geese weighing 3.6 kg, was forced to land in the Hudson River after losing its two engines in New York in 2009.

The level of risk related to collisions derives from various factors related to the geographical location of the airport, the attractiveness of the site for birds and the density of air traffic. The correct identification of these hazards brings benefits for safety management. Our study area, which is located in the middle of homes built without complying with the land-use plan, close to a rubbish dump, a rainwater collector and an open market selling various products, is exposed to the risk of aircraft colliding with birds, It is in this context that the civil liability of the state arises, as it has a duty to supervise and control the possible sources of attraction to fauna within the Airport Security Area (ASA), as well as in the areas around it.

It is on the basis of these assumptions that this dissertation seeks to discuss the problem of the risk of aviation accidents in Mozambican airspace caused by birds, with a view to understanding which and how natural and man-made factors contribute to or influence the occurrence of aircraft collisions with birds in Mozambican airspace, specifically at Maputo International Airport.

## 0.1 Problematization.

Among the various types of risk are those arising from the use of the same space by aircraft and wildlife, which is probably the most pressing issue facing airports worldwide, according to the Manager of the Airport Security Division of the *Federal Aviation Administration* (*FAA*).

Although not one of the main causes of accidents involving aircraft, aircraft collisions with birds (*Bird strikes*) represent a serious threat to civil aviation safety, having been responsible for the crash of several hundred aircraft worldwide (BLACKWELL et all. 2008, DOLBEER et all. 2016, THORPE 1996, 2003, 2012, 2016). In a recent review, THORPE (2016) counted 28 commercial aircraft and 19 executive jets that were destroyed or suffered irreparable damage due to collisions with birds, and whose accidents resulted in the deaths of more than 450 people.

Over the years, accidents resulting from aircraft collisions with birds have become increasingly frequent and intense due to various factors, such as the increase in the population of certain species of birds, the increase in bird attractions in areas close to airports, the increase in the number and size of aircraft, and the increased use of quieter and faster aircraft engines (SODHI, 2002; HESS et all, 2010). These accidents or incidents generated by collisions between aircraft and birds have caused significant financial damage and loss of human life.

It is believed that incidents and accidents resulting from aircraft collisions with birds are a direct consequence of the activities carried out at the airport and the occupation of the area around it. Air safety has increasingly become a matter of great concern, but not all the dangers that are present are known to society, as is the case with the occurrence of birds which, unfortunately, is present at almost all airports (ARP's) in the country.

The danger of birds occurring at airports tends to increase as a result of the increase in the number of flights, urban growth and a whole series of factors, indicating the need to draw up an effective work plan to try to reduce bird hotspots around airports. The presence of birds at an airport can be attributed to a number of factors and are usually related to the search for food, shelter, security, the presence of water and resting areas (ICAO, 1978). It can be seen that the potential risk of collision with birds increases as a result of the number of hotspots, i.e. anthropogenic activities, which favor the presence of birds.

The problem of collisions between aircraft and birds can be related to the ecological imbalance caused by urban growth without a proper plan, which in turn can have consequences for the population and the environment. Disordered population growth, inefficient or almost non-existent solid waste collection systems, poor basic sanitation conditions and the enormous amount of waste produced every day are the conditions experienced in the neighborhoods around Maputo International Airport.

This issue seeks to answer the question of how the prevention system inherent in avian risk, through the actions of those involved in the Mozambican aviation system, has been improving in the face of the increase in cases involving air collisions. Thus, the starting question is as follows:

✓ *To what extent do the socio-environmental conditions of the airport space influence the occurrence of aircraft and bird collisions at Maputo International Airport?*

### 0.2 Justification.

Aerodromes provide a source of food, nesting, resting and often feeding grounds for bird species which, when present, can pose a risk to aircraft operations. Fauna risk management should therefore be a routine activity and included in the operational safety management of any airport operating commercial aviation.

Today, the environment and flight safety are issues of great interest to the government and society, and are receiving a great deal of attention and attention from air transport users and information agencies.

Maputo International Airport was chosen as the field of research because it is the largest airport in the country, certified to the international standards and norms of the International Civil Aviation Organization (ICAO), with a high level of compliance required. It is also a point of entry for international traffic, making it possible to make connections with the world, receiving large aircraft that are involved in collisions with birds, thus constituting a danger to the people on board the aircraft and the material and moral damage caused by the collision.

Around Maputo International Airport, there is dense urban occupation, due to the population's desire to enjoy the benefits that the service infrastructure provides and the consequent poor management of solid waste, which can lead to the appearance of birds, and the existence of the Hulene dump, which is by the way the largest in Maputo Municipality. However, the airport brings with it a series of safety restrictions that are often forgotten or unknown by the government and municipal authorities and by society, only receiving attention when accidents occur, making national and international headlines.

Municipal authorities do not normally exercise effective control over land use and occupation, allowing activities to be set up that are often irregular and unfavorable to the proper functioning of the airport. These illegally permitted activities and the environmental factors and conditions in general, make it possible for birds to appear and cause aviation accidents and incidents, which are the subject of this study. Hence the need to establish clearer and stricter provisions in the legislation to enforce the limits set by the International Civil Aviation Organization on the occupation and use of land around airports, holding the agents responsible, whether public or private.

This research has a scientific contribution because it can help to understand the extent to which environmental conditions, such as climate, vegetation, geographical location, can affect the frequency and severity of aviation accidents caused by birds and provide relevant information to improve measures to prevent and mitigate these incidents and accidents, by designing intervention programs in order to reduce exposure to the risk of collision between birds and aircraft at airports in Mozambique, in particular at Maputo International Airport (AIM).

### 0.3 Research questions

a)      To what extent do environmental conditions favor the occurrence of collisions between birds and aircraft at Maputo International Airport?

b)      What control strategies and techniques can be taken to reduce the factors that attract birds to the airport and its surroundings?

c)      What is the state's civil liability in the event of a collision between aircraft and wildlife at Maputo International Airport?

d)      What are the main actions taken to eliminate this problem?

e)      Which measures give the best results in preventing avian danger?

### 0.4 Objectives

The main aim of this research is to help minimize the risk of bird accidents at Maputo International Airport. It should be noted that this problem is very complex in that it involves human, ecological, technical and legal components, making Maputo International Airport (AIM) a case study. In order to carry out this research, it is necessary to carry out a survey of the activities that could be attracting birds in the area surrounding AIM, as well as an analysis of the geographical context in which Maputo International Airport is located and the risks that this context poses in terms of attracting birds.

By analyzing the concepts of risk prevention and mitigation, as well as interacting with the bodies responsible for investigating and preventing aviation accidents in Mozambique, the aim is to identify those legally responsible, their spheres of action and their interests in preventing avian risk. Once identified, the aim is to compare their spheres of responsibility with their real interests in mitigating the problem.

The aim is therefore to generate knowledge and compile information on avian hazards at Maputo International Airport, so that those responsible for preventing and carrying out mitigating actions can be effective in reducing collisions with birds and their consequences.

### 0.4.1. General objective

To analyze the socio-environmental factors that influence the occurrence of collisions between aircraft and birds at Maputo International Airport between 2016 and 2021.

### 0.4.2 Specific objectives

In order to achieve the general objective, the following specific objectives were defined:

✓ Characterize the study area, Maputo International Airport (and surrounding area)
✓ Describe the air accidents that occurred in the period 2016-2021;
✓ Identify the causes of air accidents between 2016-2021, highlighting those related to birds
✓ Relating accidents to the socio-environmental conditions of the airport area

✓ Propose the revision of national legislation in order to introduce measures to prevent the risk of collisions between aircraft and birds.

### 0.5 Hypotheses

It is assumed that the presence of animals in the vicinity and especially inside the airport is related to the anthropic activities carried out around Maputo International Airport, such as the existence of a market in the area close to the airport, as well as the existence of a garbage dump "Lixeira de Hulene" in a nearby neighborhood, which are one of the main causes of the proliferation of birds that pose a risk to aviation.

The evolution of airport infrastructure, with trends towards larger airports, and the density of the population around the airport site promote an attractive environment for wildlife. According to CLEARY and DOLBEER (2005), in addition to the main attraction factors, the commercial activities typical of the airport environment, such as restaurants and commissaries, when carried out without concern for the destination of the organic waste generated, contribute to the increase of this waste in airport areas, with a consequent increase in the population of birds that are attracted.

### 0.6 Methodological procedures

Scientific work should be based on an appropriate methodology that allows us to find appropriate answers to the starting question that has been identified. This section is therefore entirely dedicated to the methodological framework of this dissertation, with a view to better understanding the chosen method. It explains how the literature was reviewed, how the questionnaire was drawn up, how the sample was defined, the interview model and how the data was analyzed.

### 0.6.1 Type of research.

This work is characterized as qualitative-quantitative research. Gamboa (1995) points out that the quantitative approach is carried out through the use of applied quantification in the collection and processing of data, using statistical procedures. According to Denzin and Lincoln (2006), the quantitative approach is used in order to isolate causes and effects, measure and quantify phenomena, while qualitative research is subjective, as it aims to seek the perceptions of those interviewed in relation to the phenomenon studied. For this study, qualitative research had the function of seeking out what the respondents or interviewees perceive about the risk of collisions between aircraft and fauna, as well as seeking possible solutions for minimizing it. It is also classified as exploratory and documentary. Secondary data from books, scientific articles, legislation and the database of the Mozambique Airports Company (ADM) and Maputo International Airport (AIM) were analyzed.

### 0.6.2 Research methods and data collection techniques

In order to carry out this research, a number of research methods and techniques were adopted, including bibliographical and documentary research, the statistical method and the cartographic method. Fieldwork was also carried out where, in addition to direct observation, interviews were conducted and surveys were carried out.

Bibliographic research consisted of analyzing books, articles, specialized journals, manuals on flight safety and avian danger at a global level as well as at a national level. For this research, the following stand out among the various researchers who deal with bird accidents: VILLAREAL, 2008, MATIJACA, 2003, PEREIRA, 2008 and SKOWO et al, 2003.

Document analysis also focuses on the researcher's perspective and involves searching for and reading written documents that are a good source of information (Bogdan and Biklen, 1994). This research used, among other documents, the Constitution of the Republic of Mozambique, the Civil Code of Mozambique, the WMD environmental management manual, the Civil Aviation Regulations of Mozambique (MOZ-CAR Part 121, 139, 12), International Air Services Transit Agreements, the Civil Aviation Law No. 5/2016 of June 16, Decree No. 73/2009 of December 15, ICAO Annex 14, ICAO publications Doc 9159, Doc 91859, Doc 98059, Doc 9859 and Doc 9139, among other documents.° 73/2009, of December 15, ICAO Annex 14, ICAO publications Doc 9157, Doc 9184, Doc 9859, Doc 9803, Doc 9859 and Doc 9137, among other reference documents in the area of preventing aircraft collisions with birds and other fauna.

The cartographic method in this research was useful in that it helped to draw up the different maps related to the geographical location of the study area, which is Maputo International Airport. It was also useful in spatializing the Airport Security Area (ASA), as well as in spatializing the possible bird hotspots in the areas surrounding the airport.

According to Bogdan & Biklen (1994), Tuckman (2002) and Quivy & Campenheoudt (2003), there are three data collection techniques that can be used as sources of information in qualitative and quantitative research: observation, surveys, which can be oral - interviews - or written - questionnaires, and document analysis. The fact that the researcher uses different methods to collect data allows him to use various perspectives on the same situation, as well as to obtain information of different kinds and then make comparisons between the various pieces of information, thus triangulating the information obtained (Igea 1995). In this way, triangulation is a process that makes it possible to avoid threats to the internal validity inherent in the way research data is collected.

As already mentioned, this research used interviews, questionnaires and direct observation. According to Bogdan & Biklen (1994), qualitative research focuses on understanding problems, investigating what is "behind" certain behaviors, attitudes or beliefs. There is no concern about the size of the sample or the generalizability of the results, and there is no question of the validity and reliability of the instruments.

For any act of research, it is always necessary to think about ways of collecting the information that the research itself will provide, for this research the techniques to be used will be the following:

Techniques based on observation are centered on the researcher's perspective, in which he observes the phenomenon under study directly and in person. This technique has been underway since 2020, where the main objective is to directly and indirectly observe the environment surrounding Maputo International Airport in order to understand whether or not there are bird attractors in the study area, the same process took place until the end of the work.

❖ **Techniques based on questionnaires and interviews**

It will be based on a survey made up of open and closed questions. This instrument will be applied to employees working in the area of environmental management as well as employees of the Environmental Management Unit (UGA), which is responsible for wildlife management, the interview will be applied to the representative of the IACM as well as the representative of the ADM linked to wildlife management.

❖ **Semi-structured interviews**

Six in-depth interviews were carried out, one with the Airport Security Manager who is responsible for wildlife management; one Pilot Commander who has witnessed an accident involving aircraft and birds; and four employees from the Environmental Management Unit (UGA) office.

The interviews made it possible to find out what each employee's tasks are, how the activities within the department are organized and how they are carried out. The interviews will be carried out individually, during the month of March and the first half of April, in a calm and comfortable environment at the company's headquarters. The interviewees will be subject to authorized recordings for later analysis. The interviews will use a script of open and clear questions, and according to the objectives of the study, "to define the most relevant dimensions of one or more attitudes" (MOREIRA, 1994:135), which condition the leaders of this organization in taking certain measures on civil liability.

❖ **Questionnaire surveys:**

In order to obtain data to achieve the objectives of this study, we selected the questionnaire survey technique. Essentially, it is a technique commonly used in research that presupposes the quantitative analysis of data since its structure is standardized, both in the wording of the questions and in their order (Borg and Gall, 2002). In addition to the reasons given, others have contributed to this being the most appropriate technique for achieving the objectives of this study. It should be noted that in order for this technique to be effective, the surveys will be hand-delivered so that the respondents can answer them.

### 0.6.3 Sample definition

The type of sampling is intentional non-probability. Non-probability sampling is where the selection of elements of the population to make up the sample depends at least in part on the judgment of the researcher or interviewer in the field. There is no known *chance* that any element of the population will be part of the sample" (MATTAR, 2001). Purposive sampling, also known as judgment sampling, is part of the group of non-probability samples. It involves greater participation by the researcher in choosing the elements of the population that will make up the sample.

For this research, a sample was defined consisting of 20 elements, all 271 ADM employees linked to the area of air transport, as well as the Airport Security sector, the Fauna Management Unit at the airport under study. The people chosen for this research had between 5 and 10 years or more as ADM employees, taking into account that the time horizon of the research is from 2016 to 2022, it was understood that only individuals who have been working at AIM for 5 or more years could or could provide the collaboration that was needed in this research, as it is assumed that they have knowledge of the different dynamics that exist in the study area.

**0.6.4 Data processing.**

Content analysis is the technique adopted for the process of processing data                           in                           order                           to data in order to transform it into enlightening information. Content analysis is basically understood, according to Bardin's (1995) definition, as a set of techniques for analyzing communications, using systematic and objective procedures to describe the content of messages.

The results of the survey will be analyzed in total or divided by group of respondents or interviewees based on the statistical tools *Spss 22.0* for their statistical analysis and later Microsoft Excel 2010 *software will be* used (tables and graphs), the study within this approach will be more comprehensive with the purpose of knowledge in the form of concrete data, and objectives.

**0.6.5 Method for assessing bird risk at the airport.**

Bearing in mind that collisions between birds and aircraft are inevitable as long as one of the two does not stop flying, the mitigation of this event at an aerodrome should aim to reduce the frequency and severity of collisions.

Considering their biological characteristics, different species pose different risks to aviation (CARTER, 2001). Thus, fauna risk management at aerodromes should prioritize efforts and human and material resources on the species most relevant to the operational safety of that region, in a site-specific approach (DOLBEER *et al.*, 2000).

In order to do this, it is important to know which species are most frequent at the aerodrome and its surroundings and which could have the most serious consequences in the event of a collision (VILLAREAL, 2008). This is possible by applying a methodology for assessing the risk of fauna on aerodromes.

ANAC, through IS 164/2015, proposes methodologies that aim to guide results for the Fauna Risk Management Program (PGRF), while CONAMA, through the publication of Resolution 466/2015, aims to authorize the management of species on aerodromes, but because it was developed with references from international aviation experts, its use is compatible in operational risk assessment.

### 0.6.5.1 Collision report query

For this particular study, in a first phase, data was collected on the history of collisions between avifauna and aircraft, data reported by ADMs at Maputo International Airport between 2016 and 2022. To this end, an annual collision index was calculated, in accordance with ICAO guidelines and literature. This index is obtained from the following Equation I:

$$I = \frac{N}{O} x1000$$

***Where:***

$I$ is the annual aerodrome crash rate;

$N$ is the total number of collisions recorded at the aerodrome during the year;

$O$ is the number of take-off and landing operations carried out at the airport during the year.

The information obtained from the index of collisions at the aerodrome made it possible to compare the years, which could help to compare the risk of fauna at Maputo International Airport with other airports. It should be noted that this index does not provide data on the risk of species associated with the aerodrome. To this end, the methodologies proposed by CONAMA Resolution No. 466/2015 were used.

### 0.6.5.2 Bird risk assessment.

To assess the risk of collisions between birds and aircraft at aerodromes, two methodologies are proposed, namely Methodology I: Collision Index in AIM and Methodology II: Quantitative assessment of problem species, where it was decided to first apply the methodology proposed by the matrix adapted from VILLAREAL (2008), referred to in this work as Methodology I, and then for species whose result was "high" or "very high", i.e. those considered to be problem species, the weighted methodology was applied, referred to as Methodology II, targeting the species that need the most attention.

The avian risk assessment can be carried out following the recommendations set out in the fauna risk matrix Table 1. This matrix can serve as an inspiration for creating a risk matrix for operational threats posed by the presence of birds in near-collision situations with aircraft.

The fauna risk matrix and the operational threat matrix for the presence of birds, in order to measure the risk that fauna and the aerodrome suffer from conflict between aircraft and birds, are available in items 3.3.11 and 3.3.1.2 of the avian risk assessment matrix (MATTOS, 2014).

According to Mattos (2014), the fauna risk matrix complies with international standards and has been divided into categories, levels ranging from zero (0) to three (3), which represent the intensity of the risk for air operations, where zero corresponds to the absence of risk and three the maximum risk.

The degree of risk (last column) is a sum of the points of the seven variables (columns A to H), which scale into four levels (High Risk, Medium Risk, Low Risk, Zero Risk), the contribution of each variable to the avian risk generated by the characteristics of the various species of animals.

**Table 1:** Fauna risk assessment matrix

| Level | Population | Pasta | Average number of individuals sighted in flocks | Presence of surveys in which the species was identified. | Period of the day (% of times sighted during the period of greatest activity at the aerodrome) | Location (% of times located in the highest risk areas) | Flight / activity ( % of times sighted in flight or intense activity) | Register (History of wildlife collision reports) | Sum |
|---|---|---|---|---|---|---|---|---|---|
| | A | B | C | D | E | F | G | H | Sum of (A - H) |
| 3 | Abundant> 50 Individuals | Very Large> 1.5 Kg | Large: ≥20 individuals. | Permanent Over 90% observation time | Permanent during the period of greatest activity at the Aerodrome: ≥ 90%. | Permanent in the highest risk areas: ≥ 90%. | ≥ 90% of the time sighted on long flights, thermals, or in intense movement around the operational area. | Collision history at the airfield in the last 5 years. | Very High Risk - from 16 to 24 points |
| 2 | Common - 20 to 50 individuals | Large - 0.75 to 1.5 Kg | Medium: ≥ 5 and <20 Individuals. | Frequent - from 60 to 90% of observation time | Frequent: ≥60% and <90%. | Frequent: ≥60% and <90%. | ≥60% e <90%. | Collision history at the airfield. | High Risk 11 to 15 points |
| 1 | Uncommon - from 10 to 20 individuals | Medium - 0.25 to 0.75 Kg | Small: ≥ 3 and < 5 individuals. | Transitional - from 30 to 60% of the | Transient: ≥30% and <60%. | Transient: ≥30% and <60% | ≥30% e <60%. | Collision history at other airfields . | Medium Risk - from 6 to 10 points |

| 0 | Rare <10 individuals | Small < 0.25 Kg | Lonely or in pairs. | observation time<br>Rare: gifts in <30% of inspections. | Rare: <30%. | Rare: <30%. | <30% | No collision history | Low Risk - from 1 to 5 points |
| --- | --- | --- | --- | --- | --- | --- | --- | --- | --- |

**Source:**     http://www.cenipa.aer.mil.br/cenipa/index.php/component/content/article/artigos-cenipa/122-matriz-de-risco-da-fauna cited by Mattos (2014).

In order to calculate the risk to fauna, the results obtained for each variable, according to Equation II.

$$\underline{Score\ do\ Risco = A + B + C + D + E + F + G + H}$$

Finally, the result obtained for risk was compared with the general risk classification matrix, as shown in Table 4.

**Table 1:** Fauna risk assessment according to overall classification score

| RISK SCORE | RISK |
|---|---|
| 16 to 24 points | Very high |
| 11 to 15 points | High |
| 6 to 10 points | Medium |
| 1 to 5 points | Bass |

**Source:** ANAC, IS nº 164/2015

Species classified as "high" and "very high" risk, i.e. problem species, require immediate measures to be implemented in order to reduce the level of risk. Species classified as "medium risk" require less urgent and/or less intense mitigation measures to be implemented. Species for which the level of risk is considered "low" imply an acceptable level of risk, requiring no new mitigation measures, but permanent monitoring.

According to Mattos (2014: 8) in this risk assessment model, a high-risk situation occurs when the abundance in the number of birds is classified as abundant, when the birds are very large, the length of stay is permanent, the previous record is of incidence at the airport, the bird's behavior is active flight and aerodrome thermal measures, the formation of flocks is considered large and the flight height is up to 30 meters. According to Table 1, this example is assigned a score of 21 points, classifying the occurrence as high risk.

An example of a medium-risk situation is one in which the Abundance in the number of birds is classified as common, the Size is large, the Length of Stay is frequent, the Previous Record is of incidents in the literature, the Behavior of the bird is short flights and active on the aerodrome, the Formation of Flocks is considered medium and the Height of Flight varies from 31 to 150 meters. According to Table 1, this example is assigned a score of 14 points, classifying the occurrence as medium risk.

Situations of low risk or tolerable risk exist, for example, when the Abundance in the number of birds is classified as uncommon, the Size is medium, the Length of Stay is transient, the Previous Record is without incident at the airport, the Behavior of the bird is perching or foraging in green areas, the Formation of Flocks is considered small and the Height of Flight is up to 30 meters. According to Table 1, this example is assigned a score of 9 points, classifying the occurrence as low risk.

Situations of zero risk, for example, are when the Abundance in the number of birds is classified as rare, the Size is small, the Length of Stay is passing, the Previous Record is without incident at the airport, the behavior of the bird is perching or foraging in green areas, the Formation of Flocks is considered solitary and the Height of Flight is up to 30 meters. According to Table 1, this example is assigned a score of 5 points, classifying the occurrence as zero risk (MATTOS, 2014:8).

**Figure 1: Flowchart of methodological procedures.**

Source: Author, 2023.

### 0.7 Structure of the work

The work is organized in 3 chapters. Chapter 1 presents a brief history of aviation safety, starting with the Chicago Convention, going through the role of ICAO, some recent aviation safety data related to air accidents and incidents, collision of aircraft and birds, the civil liability of air operators and the state in relation to accidents with aircraft and birds, it also deals with the subject related to statistical data on the problem and examples of accidents and aeronautical incidents that had the direct or indirect contribution of the risk of fauna to occur.

Chapter 2 presents the methodology and organization of the work and Chapter 3 presents the discussion of the results, the characterization of Maputo International Airport, the context in which it is inserted, the problem of the risk of fauna within the airport environment, statistical data on aircraft movements and examples of reported cases of collision with fauna at the airport, data related to the civil liability of the state in cases of collision of aircraft with fauna and finally we have the final considerations, followed by the bibliographical references used to carry out this work.

# CHAPTER 1: THEORETICAL FRAMEWORK

In view of the growth in air operations and considering the serious consequences that a collision can have, i.e. damage to human life, impacts on the environment (avifauna) and high material losses, it is necessary to monitor and develop avian risk management in proportion to this increase in aircraft in the airspace.

Thus, this chapter provides a contextualization of what avian danger is, then discusses the International Civil Aviation Organization and legislation, and finally analyzes the role or civil liability of the state in the event of accidents or incidents caused by the collision of aircraft and birds.

## 1.1 Background to the Avian Hazard

The simultaneous occupation of the airspace by birds and aircraft raises the problem of collisions involving both, which is generating growing concern among the various sectors of aviation worldwide about managing the risk to fauna, especially avifauna, both around and inside aerodromes (VILLAREAL, 2008).

Collisions between birds and aircraft have been recorded at several airports around the world and the need for safety precautions involving flights has become even more evident (MATIJACA, 2003). These collisions can have consequences ranging from the slightest, practically imperceptible to passengers, to the aircraft going out of control and crashing (PEREIRA, 2008), which generates damage, whether material, financial or even involving the loss of human life.

Hazard differs from risk in that the former represents a qualitative measure of an existing condition in the environment capable of causing loss or damage (VINCOLI, 2006) and the latter, according to Stephenson (1991), is a statistical representation of the possibility of undesirable situations occurring. The understanding of avian danger, as well as the generating factor and the search for measures to manage and reduce this risk are also relevant in Mozambique.

The term "avian danger" is an imminent risk of a collision between an aircraft and a bird or flock of birds, on the ground or in the airspace, especially during the approach, landing and take-off phases, which is made up of two variables: the probability of a collision and its severity.

The probability of a collision is the ratio between the flow of aircraft operating at a given airport and the number of birds present at the operating sites, whether on the ground or in flight. Gravity depends on the speed of the aircraft and the proportion of the bird's mass. Therefore, the higher the speed of the aircraft and the greater the weight of the bird, i.e. its mass, the more serious the consequences of this collision will be (ANAC, 2011 cited by Fernandes 2017:18). There are two types of collision that occur between aircraft and birds: the first type is related to the bird or flock of birds being swallowed by the aircraft's engines during approach, take-off and landing procedures and colliding with the aircraft's windshield.

When the collision involves ingestion into the engine, the take-off is aborted or an emergency landing is made, both of which are high-risk procedures, as they occur in one of the critical phases of a flight: heavy aircraft with a lot of fuel, minimum speed required for take-off, low altitude and close to an urban area (RAMOS, 2009).

Figure 2: Bird ingestion during take-off

Source: https://bashny.net/t/es/46036

Birds colliding with windshields can cause them to shatter, making piloting more complex due to poor visibility and the strong relative wind that enters the cockpit. It can also cause injuries to pilots due to the force of the bird's impact.

**Figure 3: Aircraft windshield after collision with bird**

Eschenfelder (2001) points out that avian risk is a risk to airport security and is caused by the presence of birds in and around the airport. As civil aviation activity continues to expand, wildlife populations have also grown in the vicinity of airfields and collisions between these animals and aircraft are on the increase.

The presence of animals in the vicinity and especially within the airport can be attributed to various factors, as already mentioned, usually related to the availability of food, water, shelter, security, nesting and resting areas. The presence of a group of birds can also be directly related to the local vegetation (SKOWO et all, 2003). Most of the time, the airport community is unaware of the dangers of keeping domestic animals in a movement area, which also influences operational safety. Commercial activities at the airport, such as restaurants, carried out without due concern for the final destination of the organic waste generated, contribute greatly to the increase in the bird population.

### 1.2 Mitigation Measures against Avian Hazards

In October 2006, the International Bird Hazard Committee (IBSC) published a document containing recommended practices for bird and animal control at airports entitled *'Standards for Aerodrome Bird/Wildlife Control'*.

This document contains the main measures to combat avian danger used around the world, along with help in creating a specific program for each aerodrome. The document cites nine standards that must be followed in order to achieve a successful avian hazard control program. These are: (*International Birdstrike Committee*, 2006)

*1. Firstly, a senior member of the airport administration should be responsible for implementing an Avian Hazard Control Program, including habitat management and active avian hazard control;*

*2. The airport should conduct a survey of its property in search of resources that could be attractive to birds. These resources should be well identified and a plan made for their management with the aim of reducing their quantity, or preventing bird access to them where practicable. This plan should be well described and highlighted for future analysis;*

*3. A team trained to chase away birds must be present in the airfield at least 15 minutes before each landing or take-off, and in the case of busy airports they must be on standby all day. This team should only be concerned with birds and even at night, teams should patrol the airport to identify any danger;*

*4. The teams must have the necessary equipment to scare away or capture the birds that live there, taking into account the species, the number of birds and the area to be controlled;*

*5. Control teams should keep constant notes on: areas that have been patrolled; number, species and location of birds sighted; actions taken to scare them away; results of the actions;*

*6. Accidents should be divided into 3 categories: Confirmed Accidents; Unconfirmed Accidents and Confirmed Major Accidents;*

*7. Airports should establish some mechanism to ensure that all bird accidents in their region are reported. The species involved should also be identified whenever possible. This data should be passed on to the national avian hazard agency, which should pass the information on to ICAO every year;*

*8. Airports must carry out a formal assessment of the risks posed by the avian hazard in order to evaluate the effectiveness of the control program;*

*9. Airports must ensure that the implementation of bird attracting activities is prohibited within an area comprising a circle with a radius of 13 km.*

## 1.3 Regulatory framework

The presence of birds is a threat to flight safety. In order to mitigate this risk, international and national laws, standards and rules have been created, addressing parameters and recommendations to be followed, as follows:

The ICAO (International Civil Aviation Organization) was created in 1944 to serve as the basis and standard for aviation activity worldwide, with the aim of conducting aviation safely. From 1965, the ICAO began to monitor collisions between aircraft and birds through reports. In 1980, an automated system was created for storing these reports, IBIS (ICAO *bird strike information system*). According to its latest survey, by 2015 this system had recorded 97,751 collisions worldwide.

ICAO began collecting data on avian danger in 1965, and in 1980 programmed the *Bird Strike Information System* (IBIS), a system for collecting and disseminating data on aircraft collisions with birds (MADEIREA, MARTOS, 2013).

From an international perspective, the main ICAO document that regulates and provides recommendations on this subject is Annex 14 (Manual on the ICAO Bird Strike Information System - IBIS Doc 9332). This document contains three practical recommendations, which are:

1)	This refers to the reporting and storage of data on collisions between aircraft and birds, in which the damage caused must be analyzed through a national procedure for collecting information from operators: airline and airport management.

2)	The airport authority must seek preventive measures to reduce the number of birds that pose a risk to aircraft operators around the airport.

3)	Proper control and storage of solid waste (garbage dumps) generated by humans in order to avoid attracting birds to the airfield area.

4)	Mozambique, being a member of ICAO and following the recommendations regarding the control and reporting of accidents related to collisions between aircraft and birds, and based on the recommendations established by ICAO, created and published Civil Aviation Law No. 5/16 of June 14, which establishes a regulatory framework for the area of civil aviation, as a way of guaranteeing the protection of the public interest and national air safety, ensuring compliance with international air safety standards in all civil aviation operations. In this context, in Chapter IX Investigation of Accidents and Aeronautical Incidents, Article 69 on accidents and aeronautical incidents, recommends the establishment of standards and procedures for the investigation of accidents and aeronautical incidents in specific regulations, in accordance with the provisions of the Convention on International Civil Aviation. To this end, "MOZ-CAR - PART 12" (*Mozambique Civil Aviation Regulation*) was created by Ministerial Order No. 117/2011, of May 3, whose main objective is the investigation of accidents and incidents involving civil aircraft, with a view to promoting the prevention of accidents and incidents involving aircraft, including accidents or incidents related to collisions between birds and aircraft. MOZCAR PART 139 Subpart VII on Bird Hazard Management establishes procedures for managing fauna and reducing the risk of birds on the aerodrome. These procedures only apply to managing the occurrence of birds within the airport perimeter.

For its part, ADM, E.P. is aware of the harmful effects that can be caused during its activities. In order to reduce these effects, the company created the *Environmental Management Manual in* 2011, under the tutelage of the Environmental Management Unit (UGA). This office is responsible for managing airport infrastructures, solid waste, liquid waste, air management, noise management, water management, soil management (erosion control), nature maintenance (preservation of animal and plant habitats) and fauna management on the airport site.

According to PNFTCSAC (2000:10)[5] "Security restricted area" means those areas on the airside of an aerodrome identified as risk areas where, in addition to access control, other security controls are carried out. As a rule, these areas include, in particular, all the commercial aviation passenger departure areas between the screening points and the aircraft, the traffic area, the baggage screening areas, including the apron and the areas where baggage and cargo are placed after screening and cargo dispatch, and the airside parts of the cargo terminal, post office and aircraft cleaning and *catering* services.

---

[5] National Training and Certification Program

**Figure 4: Representative model of bird strike prevention and related entities**

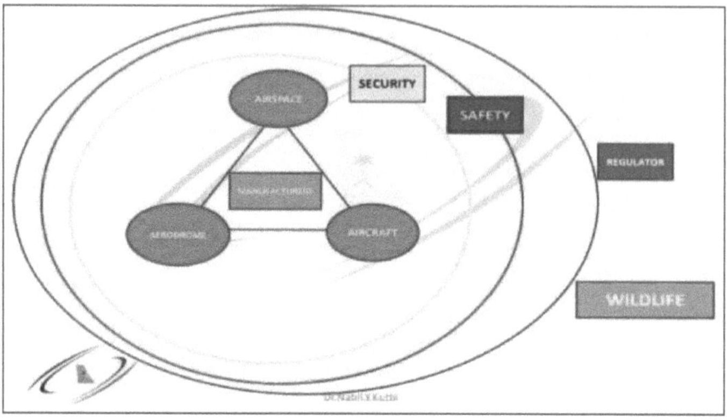

*Source:* PNFTCSAC (2000:10)

## 1.4 Management and control of Avian Hazards at Airports

In a broader context, it must be considered that the presence of an airport in a given region, as well as the definition of flight routes, must be planned taking into account that the urban space encompasses an ecological system with animal life established before the presence of the airport.

The risk of fauna does not arise by chance, but is linked to the development of the city, the transformation of the environment and the landscape. Hence, all airport activity must be planned logistically and strategically in order to foresee the consequences of urban interference in a given geographical region. Air operations near airports must be organized in such a way that each interference and its respective implications are surrounded, in order to guarantee flight safety.

The threat of avian danger is universal. Birds do not respect any space or airport boundaries, flight phases, type of aircraft, season and crew experience. Yet the solution lies in all these areas.

Reducing the risk to birds depends on many factors that are constantly changing, where civil aviation authorities, air transport companies, military aviation and crew members can make important contributions to preservation. But for these endeavors to have a positive effect, efforts need to be coordinated, since the magnitude and dynamics of the problem require a systemic and continuous approach to efforts (MENDONÇA, 2008).

Birds are attracted to airports for various reasons, all related to survival, but these basic needs increase the risk of accidents caused by avian danger unless an efficient wildlife management program is implemented at the airport.

The IACM is the regulatory body for civil aviation in Mozambique. Created by law 41/01, of December 11, 2001, the IACM is a public law entity with financial and functional administrative autonomy, under the supervision of the Minister who oversees the area of civil aviation, this body strives to mitigate risks, with inspection, verification of procedures and measurement and observation of mitigating measures.

The IACM's efforts in coordination with the Airport Fauna Control Unit (UCFA[6] ) are focused on the systematic inspection of airport infrastructures at national level and in particular at Maputo International Airport, where the fauna hazard management procedures are analyzed, as well as the conditions of the airport enclosure with regard to bird activities.

One of the IACM's important activities is related to the airport operational certification process, in which the fauna hazard management program is an integral part of the process. In addition, the mitigating measures put in place by airport operators are observed, such as the management and

---

[6] Airport Fauna Control Unit

control of local fauna, maintenance of green areas, control and elimination of fauna hotspots around airports and airfields, among other actions.

In this sense, the IACM plays an important role in supervising and monitoring the work carried out inside airports. Among the various roles played by the IACM in monitoring and regulating air operations, the safety of these is assumed to be fundamental.

According to the MOZ-CAR Part 139 Civil Aviation Regulations of Mozambique, whose purpose is to establish the requirements for the construction expansion modification, certification and licensing of public and private aerodromes, in *Subpart VII - Animal Hazard Management,* paragraph 139.7.3 and 139.7.4, establish procedures to address hazards to aircraft operations arising from the presence of birds on the flight paths at the airport or animals in the movement area, which should include: procedure for assessing existing hazards; and, procedure for implementing programs to control vegetation cover, fauna, and anthropogenic activities.

### 1.4.1 Wildlife management (139.7.3)

1) The operator, in consultation with the Authority responsible for the conservation of fauna, must take the necessary measures to control the dangers of the presence of birds and animals on the aerodrome;
2) The Operator must ensure that the procedures for dealing with the danger arising from the presence of birds and animals in aircraft operations and the aircraft movement area are guaranteed;
3) The Aerodrome Animal Management Plan must be approved by the IACM and must form part of the Aerodrome Manual.

## 1.4.2 Reducing the risk of birds on the aerodrome (139.7.4)

1. In consultation with the Authority responsible for the conservation of fauna, the operator must take all reasonable measures to minimize the risks associated with bird hazards at the Aerodrome.

2. The operator must take concrete measures to control the habitat and disperse birds around the Aerodrome that constitute a potential danger to aircraft operations.

3. The danger of an accident involving birds on or near an aerodrome must be assessed by means of:

✓ Procedure established for recording and reporting bird accidents with aircraft;

✓ Gathering information from aircraft operators and aerodrome personnel or any other person about the presence of birds around the aerodrome that constitute a potential danger to aircraft operations.

4. The operator shall prepare a report on the danger caused by birds at an aerodrome using the information collected under the provisions of the subparagraph.

5. The operator must send the bird hazard report to the IACM for transmission to the International Civil Aviation Organization (ICAO) for inclusion in the Bird Accident Information System database.

6. The operator must take measures to eliminate or prevent the creation of waste disposal sites or any other source of waste that could constitute a source of attraction for birds on or near an Aerodrome, unless an appropriate aeronautical study indicates that the dumps are not likely to create conditions favorable to a bird hazard.

7. The aerodrome operator must establish and manage a bird hazard control unit.

## 1.5 Fauna risk management program (PGRF).

A PGRF, which responds to risk in the sense of mitigating or reducing the risk of collision between birds and aircraft, formally structures procedures with assigned responsibilities and procedures that are integrated into the operational routine (OLIVEIRA, 2017).

From the aerodrome administrator to the operators at the most basic levels have a stake in the PGRF, which considers various factors in the proactive implementation of practices.

According to IWERSEN (2018:24), what underpins a PGRF are the passive measures that must be implemented in order to resolve the risk immediately, because modifying the environment will mean mitigating the points of attraction for birds. In this sense, the focus will be on the lowest level of the pyramid to solve the problem of avian danger.

**Figure 5: Technical priority of fauna control measures at aerodromes.**

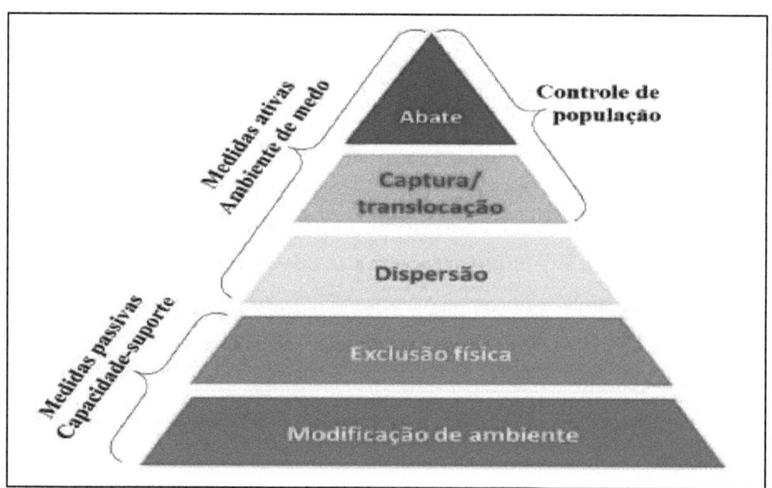

In the PGRF, some measures can be seen as strategic in controlling and managing avian risk. Recent technologies can make this possible. On a legal basis, the PGRF is organized politically, assigning responsibilities that can be verified later.

Objectives and targets are proposed considering the reality of the aerodrome and, having established the qualification and training of the teams with the appropriate resources, the entire risk context is mapped according to the reality of the aerodrome. With established routines, complete data records and the implementation of measures and strategies, efficiency is evaluated and actions are revised, if necessary.

### 1.6. State civil liability in relation to aircraft collisions with birds.

The Civil Code contains general rules on civil liability, but there are specific provisions in sectoral legislation that contain particular rules on how civil liability will operate in the situations it describes. One of these sources that contains specific rules is Law No. 5/2016 of June 14, which approves the Civil Aviation Law of Mozambique, which contains special provisions on civil liability in domestic and international air transport.

As a rule, the Civil Code establishes that whoever is obliged to repair damage must restore the situation that would have existed if the damage had not occurred. However, the Civil Aviation Law contains its own provisions, applicable to situations such as anomalies, missed flights, overbooking and damage to luggage and goods, establishing different compensation for each of these situations depending on the case.

Generally speaking, transportation has always been treated differently by national legislation, especially air transportation, because of the strong theory of risk, which assigns liability for damages to the person who carries out the activity at risk, where it gives the operator of the activity the duty to observe both the profits of that activity and its risks. Thus, the protection of the passenger, who hands over his most precious possession, his life, to a businessman or to the state itself, in order to be transported (CAVALIERI FILHO, 2005:316).

Accidents and incidents in the air are a source of enormous concern and fear for everyone. However, air travel is one of the safest modes of transportation in the world, second only to elevators when it comes to operational safety. The probability of a fatal air accident is minimal, at one in 8 million; whereas in the case of other means of transportation such as cars, the *chance of an* accident is one in 18,800. Compared to air transport, car transport has much more alarming statistics. However, the repercussions of an air accident on society are much greater (MARTINS, 2015 *apud* MACHADO, 2017). Looking at the ICAO's historical data on aviation accidents, it can be concluded that there has been a reduction in air accidents. Even so, the occurrence of this type of event always causes great trauma to the population (CAVALCANTI, 2002).

However, the ICAO recognizes that the airport operator has limited efficiency, since birds easily pass through the approach and take-off paths, where other authorities must collaborate to prevent avian risk. As can be seen in Annex 14 - Aerodromes cited:

> *The appropriate authority shall act to eliminate or to prevent the establishment of solid waste disposal sites or any other source that may attract wildlife to the aerodrome or its vicinity, unless appropriate assessment indicates that such sites are unlikely to create conditions that lead to wildlife problems. Where elimination of existing sites is not possible, the appropriate authority shall ensure that the risks caused to*

*aviation by such sites have been assessed and reduced to the lowest practicable risk condition (ICAO, 2009:9-10).*

In Mozambique, most of the airports are located in the middle of urban centers, where it is possible to find various bird attractors, and due to the activities that take place around these airports, garbage dumps and other bird attractors end up appearing, thus infringing environmental and aviation legislation simultaneously, In this war of responsibility, it should be noted that the area of responsibility of the airport operator is smaller than the area of responsibility of the municipal and governmental authorities around the airport.

As you can see, the area belonging to the airport is smaller in relation to the area around it, which means that there is no logic in considering that it is the airport operator's responsibility to control all the hotspots in the area of interest, and therefore the birds that are collided with by aircraft operating at the aerodrome.

The state, which is largely responsible for environmental management (Ministry of Land and Environment), for air transport services through the IACM, as well as for defining urban development and environmental policies (Municipalities), whose main objective is to order the development of the functionality of the space and guarantee the well-being of its inhabitants (Law 19/2007 of July 18), Since this state is the guarantor of the functionality of the space around the airport enclosures, it is possible to conclude that the state has clear responsibility for Air Accidents resulting from collisions with birds, especially when the collided birds result from the creation of bird hotspots around the airport enclosure.

It is important to note that the state's civil liability in the event of a collision with birds will only generate compensation for the damage, in the first instance, if it is requested by the victim and, ultimately, by the airline that has already borne the cost of repairing the damage to its aircraft and to the people who may have already been harmed as a result of the event, and then, when it can be established that there is a causal link, i.e. that the birds that collided with the aircraft arose from the existence of hotspots, as a result of the state's mismanagement of the areas around the airport.

**1.7 Identifying bird attractions in the airport area**

According to ANAC (2016) cited by Costa (2017), environments that can serve as an attraction for birdlife are those that promote the availability of shelter, water and food. Thus, in airport environments, the extensive areas covered in grass, water retention sites, fruit trees and built infrastructures that serve as perches stand out.

The grass at the airport: The grass at airports is considered to be a source of attraction for fauna, since when it is too tall, it provides shelter for fauna, while grass that is too low can mean that it is a suitable nesting area for various species, such as the red kite. In addition, lawns can be made up of different species of grass, some of which produce fruit or seeds, and in this type of undergrowth it is common to find insects and grains that serve as food for various birds. Ultimately, the mowing process generates vegetation clippings which are also attractive to wildlife.

Other areas of vegetation: Areas with forest formations and fruit trees, even if isolated, are considered attractive due to the large number of birds that use them as perches, feeding grounds, roosting and nesting areas.

❖ Drainage systems and wetlands: Drainage channels, when not properly maintained, as well as other areas susceptible to flooding, can accumulate water during the rainy season, attracting birds for feeding and waterfowl.

❖ Roosts: The presence of trees (live and dead) in the aerodrome's operational area, air navigation instruments and antennas, lampposts and the airport's own buildings and fences are an attraction for various species as they provide suitable environments for staying and resting.

❖ Attractive fauna: There are species which, although they don't directly interfere with aviation operational safety, act as an attractive focus for problem species. This is the case with species of herpetofauna and insect colonies such as bees, ants and termites, as well as other invertebrates. During the breeding season, insects fly back and forth, increasing the potential for attracting birds. Fish that serve as food for birds and cattle that provide food for birds while grazing are also considered to be attractive fauna. In addition, eggs and chicks can also serve as food for other species and are also characterized as such. Finally, the death of animals can attract concentrations of vultures and other birds with detritivorous behavior if the carcasses are not collected before the organic matter begins to putrefy.

❖ Solid waste and effluent treatment plants: Liquid effluents and solid waste, if poorly managed, because they represent a source of food for fauna, can become an attraction for fauna in search of food.

❖ Protection systems: Although this study does not include a risk assessment for mammal species, it is important to point out that flaws in operational and property fences and the absence of drainage channel grating can allow terrestrial animals access to the airport's operational area, thus posing a risk to operations.

Businesses considered to be attracting fauna are those which, during the production process, can provide food for birds. Table 1 describes the analysis criteria for issuing CENIPA's technical opinion, according to the potential fauna attraction of each activity, depending on its respective distance from the ASA: 5 km, 10 km and 20 km.

**Table 2**: Special restriction related to the distance from the center of the largest runway of the aerodrome.

| Type of Activity | Potential for attracting fauna | Special restriction related to the distance from the center of the aerodrome's longest runway | | | |
| --- | --- | --- | --- | --- | --- |
| | | New development (LP and LI) | | Enterprise in operation | |
| | | *Up to 5 km* | *Between 5 and 10 km* | *Above 10 to 20 km* | *Over 20 km* |
| **Slaughterhouse** | Very high | Prohibition | Suitability | Suitability | Suitability |
| **Extensive farming** | High | Prohibition | Suitability | Suitability | Suitability |
| **Aquaculture or fish processing (open)** | Moderate | Suitability | Suitability | Suitability | Suitability |
| **Controlled landfill** | High | Prohibition | | | |
| **Landfill** | High | Prohibition | | Suitability | Suitability |
| **Dams (water mirror creation)** | High | Prohibition | Suitability | Suitability | Suitability |
| **Raising beef animals** | Moderate | Suitability | Suitability | Suitability | Suitability |
| Tanneries | Very high | Prohibition | Prohibition | Prohibition | Prohibition |
| **Open dumping of solid waste (dump)** | Very high | Prohibition | | | |
| **Solid waste transshipment station** | Moderate | Suitability | Suitability | Suitability | Suitability |
| **Sewage or water treatment plant (WTP)** | High | Suitability | Suitability | Suitability | Suitability |
| **Street markets (foodstuffs)** | Moderate | Suitability | Suitability | Suitability | Suitability |
| **Food processing industry (feed, etc.)** | Moderate | Suitability | Suitability | Suitability | Suitability |
| **Markets** | Moderate | Suitability | Suitability | Suitability | Suitability |

| | | | | | |
|---|---|---|---|---|---|
| **Residences** | Moderate | Suitabilit y | Suitabilit y | Suitability | Suitability |
| **Silos and other food storage buildings** | Moderate | Suitabilit y | Suitabilit y | Suitability | Suitability |

Source: Doc. 9137, ICAO 2012

As an example described in Table 2, we can mention the controlled landfill (daily collection - inert material) and open-air solid waste disposal (dump), which have a very high potential for attracting fauna and it is forbidden to set up this enterprise throughout the ASA, i.e. up to a radius of more than 20 km. Another example is the sanitary landfill (daily collection - inert material), which is also considered to have a very high potential for attracting fauna and is prohibited from being implemented up to 10 km from the aerodrome.

# CHAPTER 2. PHYSICAL-NATURAL AND SOCIO-ECONOMIC CHARACTERIZATION OF THE MAPUTO INTERNATIONAL AIRPORT AREA

The main focus of this chapter is to present the physical-geographical and socio-economic characteristics of the study area, as well as its geographical location.

## 2.1 A brief history of the city of Maputo

Maputo is Mozambique's capital and largest city. It is also the country's main financial, corporate and commercial center. It is located on the western shore of Maputo Bay, in the far south of the country, close to the border with South Africa and the border with Eswatini and, therefore, the triple border of the three countries.

Founded in the 16th century, it served as the main Portuguese trading post at that point in the Indian Ocean, thanks to its privileged bay, and in 1898 it became the capital of the colony; until March 13, 1976, the city was called "Lourenço Marques" in honor of the Portuguese explorer of the same name.

**Figure 6: Maputo City**

Source:      Panorama      do      Centro      de      Maputo, (https://pt.wikipedia.org/wiki/Maputo)

The city is administratively a municipality with an elected government and has also had the status of a province since 1980. It should not be confused with the province of Maputo, which occupies the southernmost part of Mozambican territory, except for the city of Maputo.

The municipality has an area of around 300 square kilometers and a population of 1,088,449 (2017 Census). Its metropolitan area, which includes the municipality of Matola and the districts of Boane and Marracuene, has a population of 3 158 465 inhabitants (INE, 2019).

## 2.2 Location of Maputo City

The city of Maputo is located in the south of Mozambique, west of Maputo Bay, on the Espírito Santo Estuary, where the Tembe, Umbeluzi, Matola and Infulene rivers flow. The city of Maputo has an area of 346.77 km$^2$ and borders the district of Marracuene, to the north; the municipality of Matola, to the northwest and west; the district of Boane, to the west; and the district of Matutuíne, to the south; all belonging to Maputo province. The city of Maputo is located 120 km from the border with South Africa and 80 km from the border with Eswatini (GOVERNMENT OF THE CITY OF MAPUTO, 2015).

The municipality's boundaries lie between latitudes 25° 49' 09" S and 26° 05' 23" S and longitudes 33° 00' 00" E and 32° 26' 15" E (GOVERNMENT OF THE CITY OF MAPUTO, 2015). Its population according to the 2017 census points to a slight decrease in population from the 1,111,638 recorded in the 2007 census to 1,101,170, 10468 fewer inhabitants or 0.9%. This decrease followed a weak growth of 13.2% among the 966,837 inhabitants listed in the 1997 census. Population growth between 1997 and 2007 is equivalent to 1.2% per year, half the national average of 2.4%, and has a population density of 4033 inhabitants/km$^2$ (INEM, 2017).

### 2.3 Geographical location of the Mavalane neighborhood.

Maputo is divided into seven municipal districts, which are in turn divided into neighborhoods and villages.

Our study area is an integral part of the city of Maputo, located in the KaMavota Municipal District, where to the south it borders the Urbanization District, to the west it borders the districts of Nsalene, 25 de Junho A, Bagamoio, to the east the district is bordered by the districts of Hulene "A" and to the north by the districts of Magoanine "A" and Laulane. Its population according to 2017 census data is 19,407 inhabitants (INE, 2017).

### 2.3.1 Natural physical aspects of the Mavalane neighborhood.

According to Muchangos (1994), from the point of view of natural conditions, Maputo belongs to the Southern Africa macro-region, which stretches from the Zambezi River in the north to the extreme south of the African continent. As such, all aspects related to physical characterization are closely related to that macro-region.

In terms of geomorphology, the city of Maputo is characterized by the existence of coastal plains and *plateaus*, depressions and slopes, and the predominance of sandy, alluvial and brownish-clay soils that have evolved and are susceptible to water erosion (OMBE et all, 1996 and Atlas Vol.1, 1986).

Morphologically, the city of Maputo is dominated by a coastal plain landscape, which developed from the Pleistocene onwards and presents an alternation of insensitive landforms in very small spaces. According to Barradas (1965) quoted by Muchangos (1994), five transgressions alternated with six regressions occurred throughout the southern Mozambican plain during the quarternary. These repeated changes in mean sea level are closely linked to terrestrial phases of morphogenesis. The arid climate that accompanied the transgressions favored wind activity, allowing the formation of the thick layers of sand that make up the region's most important geological substrate.

The district of Mavalane, the study area, consists of or is occupied by a predominantly sandy surface with a slightly undulating relief, with altitudes varying between 26 and 40 meters.

In terms of soils, the Mavalane district is made up of sandy soils of low fertility and low water retention, characteristic of the most prominent area of the district, i.e. the highest area (40 meters). According to the Geographical Atlas, Vol. 1 (1986), in the study area you can find a mixture of the soils described above and the evolved brownish sandy loam soils of intermediate and good fertility, and in parts thin soils occurring in areas of depressions, which are used for the production of fresh produce such as lettuce, onions, tomatoes, among other products.

In terms of climate, Maputo has a humid tropical climate, with two distinct seasons: a hot and humid season from October to March, and a dry and cool season from April to September.

The climate of this region is affected by three main factors (DNA, 2005): the position of the southern intertropical convergence zone and anticyclones in the South Atlantic and Indian Ocean and thermal depressions during the hot season.

The average annual temperature and rainfall in the city of Maputo at Mavalane station were 23°C and 813mm respectively, with January to March and October to December being the wettest months. The average relative humidity is 66.6%, with little fluctuation throughout the year. The month with the highest relative humidity is March with 71.0%, and the month with the lowest humidity is June with 63.5%. Figure 1 shows the average monthly rainfall in Maputo city at the Mavalane station during this period. (INAM, 2023)

**Figure 7: Average monthly rainfall in Maputo City at Mavalane station.**

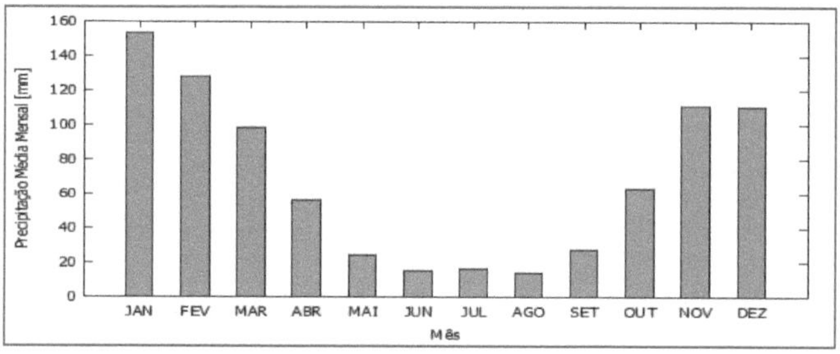

Source: INAM, 2023.

With regard to the vegetation layer in the study area, according to Muchangos (1994), the city of Maputo, looking at the rainfall regime, has an open forest, but over time this forest has been replaced by savannah-type vegetation, with strong evidence of anthropogenic influence, which is called Savannah in use.

The vegetation in the Mavalane neighborhood is poor, with the sandy soil consisting of grass and the trees that do exist are confined to the outskirts of the houses, where a more wooded area is characteristic. These vegetation characteristics can influence collisions between birds and aircraft, as birds usually look for areas with this type of vegetation to find food.

**Map 1: Map of Maputo City's Administrative Division**

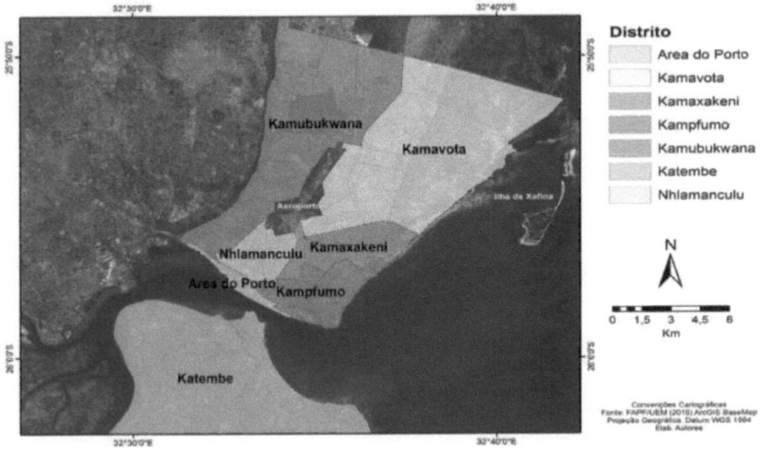

Source: CMCM, 2023.

| Distrito Municipal | Bairros |
| --- | --- |
| KaMpfumo | Central A, B, e C; Alto Maé A e b; Malhangalene A e B; Polana Cimento A e B; Coop e Sommerschield |
| KaNlhamankulo | Aeroporto A e B; Xipamanine; Minkadjuine; Unidade 7; Chamanculo A, B, C e D; Malanga e Munhuana |
| KaMaxaquene | Mafala A e b; Maxaquene A, B, C e D; Polana Caniço A e B e Urbanização |
| KaMavota | Mavalane A e B; FPLM; Hulene A e B; Ferroviário; Laulane; 3 de Fevereiro; Mahotas; Albasine e Costa do Sol |
| KaMbukwane | Bagamoyo; George Dimitrov (Benfica); Inhagoia A e B; Jardim; Luis Cabral; Magoanine; Malhazine; Nsalane; 25 de Junho A e B, e Zimpeto |
| KaTembe | Gwachene; Chale; Inguice; Ncassene e Xamissava |
| KaNyaka | Ingwane; Ribjene e Nhaquene |

Source: wickpedia, 2023

# CHAPTER 3: INFLUENCE OF ENVIRONMENTAL CONDITIONS ON THE OCCURRENCE OF AVIATION ACCIDENTS ASSOCIATED WITH BIRDS: THE CASE OF MAPUTO INTERNATIONAL AIRPORT

The purpose of this chapter is to present the results collected in the study area through the use of surveys and interviews with employees at Maputo International Airport. The first stage was to present the characterization of Maputo International Airport, followed by data related to gender, age, education, position, area of occupation, information on the occurrence of accidents related to or associated with birds, among other information.

## 3.1 Characterization of Maputo International Airport.

### 3.1.1 History of Maputo International Airport

**Figure 8: The first terminal opened in 1940**

Source: LAM website, 2023.

The first terminal was inaugurated in 1940 under the name of Lourenço Marques Airport. It was located in the building where the current Directorate General of Mozambique Airlines (LAM) now operates.

**Figure 9: The second terminal opened in 1962**

The second terminal opened in 1962, as part of Gago Coutinho - Lourenço Marques Airport, the name it had at the time.

**Figure10: Old Control Tower (1958) New Control Tower (2009)**

Gago Coutinho Airport                    Maputo International Airport

In February 2006, Aeroportos de Moçambique (ADM EP) began a project to modernize and expand the Maputo Airport. The work included the construction of a new international passenger terminal, which was inaugurated on November 12, 2010; a domestic and international cargo terminal and a new control tower, completed in 2009; and, finally, a domestic passenger terminal, to be put into operation in October 2012.

### 3.1.2 Characterization of the study area

Maputo International Airport (ARP/MA) is located 3 km N of Maputo City Center, in the district of Mavalane, at an elevation of 147 FT (feet), which corresponds to 45 meters from mean sea level, a reference temperature of 28° C and has the following geographical coordinates: 25° 55' 15.0" South Latitude and 032° 34' 24.5" East Longitude (AIP Mozambique, 2021).

**Figure 11: Location of the "Maputo International Airport" study area**

Source: Image taken from Google, (06/06/2022).

Its official address is: AEROPORTOS DE MOÇAMBIQUE, E.P.- Maputo International Airport - Alameda do Aeroporto - Caixa Postal - Cidade de Maputo-Mozambique. The airport covers an area of 14km and is bordered to the north and east by the municipal districts of Kamubukwane and KaMavota, to the east by the municipal districts of KaMavota, KaMaxaquene, KaMpfhumu and KaNyaka, and to the south by the municipal districts of Kalhamankulu, KaMpfhumu and KaTembe. Access can be via three roads (Avenida Julius Nyerere/Rua da Beira; Avenida Acordos de Lusaka; Avenida de Angola/Rua Gago Coutinho/ Avenida 19 de Outubro/ Largo da Deta).

### 3.2 Physical characteristics

Maputo International Airport (ARP/MA), which handles international traffic, is designated as an aerodrome for the entry and exit of international and domestic air traffic, where all procedures relating to migration, customs, health, agriculture, animal and plant quarantine and similar procedures are applied where traffic services are available on a regular basis.

These characteristics classify it as a Main Aerodrome according to Mozambique's AIP (Aeronautical Information Publication), a basic aviation document designed primarily to meet national and international needs for the exchange of aeronautical information.

It has two runways with a magnetic orientation of 05/23 and 10/28. The main runway, which is in use for landing and take-off, is 3660x45m and has a sidewalk strength of PCN 26F/B/X/U, and the secondary runway is 1700x45m and has a sidewalk strength of PCN 24F/B/X/U, respectively. The airport covers an area of 804.13 hectares and has a perimeter of 13,939.41 hectares.

### 3.3 Passenger Terminals

The terminal at Maputo International Airport has two modernized terminals:

### 3.3.1 Domestic Passenger Terminal

The domestic terminal has the capacity to handle 580 passengers in one hour, 300 on departure and 280 on arrival, and the infrastructure is around 13,000 square meters.

### 3.3.2 Facilities offered

The building has 14 *check-in* counters, equipped with modern passenger service technology, four entrance doors, two escalators, elevators, escalators, a toilet area, a flight information system, security equipment (X-ray, metal detectors), two boarding/departure lounges, work offices in the public space and a commercial area, thus meeting the essential needs of airlines and other service providers, such as a public catering area.

On the land side, there is an area for gardening and decorating, as well as parking for vehicles. On the air side there are exit sleeves for passengers.

### 3.3.3 International Passenger Terminal

The building has a large departure lounge prepared to handle 900,000 passengers a year and is properly equipped with modern technologies.

### 3.3.4 Facilities offered

• It has three boarding gates, 13 computerized *check-in desks, the* same number of migration desks on arrival and another seven on departure and has the capacity to handle four aircraft simultaneously, two of which are large.

• A parking lot with capacity for around 800 vehicles;

• A garbage incinerator;

• Three X-ray machines;

• Computerized check-in system;

• Personalized passenger assistance via a passenger service desk and toll-free line;

• Two bridges Passenger embarkation/disembarkation;

• It also has a new cargo terminal with its respective accesses and a parking lot for 40 vehicles;

- A control tower equipped with modern communications equipment and using state-of-the-art technology, as well as access points and a vehicle parking lot (source: ADM, Operations Directorate, 2022),

The two passenger terminals (A and B) house restaurants, banks, telephone services, post offices, ATMs, tourist information and urban cab services. Next to these terminals is a building housing the airport administration and the operational control areas of some airlines. The terminals have a large parking area for staff and passengers. The airport also includes the Mavalane Air Base, which has a new VIP Presidential pavilion and access, as well as a parking lot for around 90 vehicles; and the cargo terminals, which have all the facilities for handling cargo.

**Figure 12: Image of the International Parking Sign**

Source: Author, 2022

**Figure 13: Airside image of Passenger Terminal B**

Source: Author, 2022

### 3.4. Aircraft maintenance area

The airlines have an industrial area in the eastern sector of the airport, near Pier 28, where maintenance services are carried out in hangars specific to each type of repair, as well as buildings for administrative services.

### 3.5 Runways and Aircraft Parking

The landing and take-off areas are runway 05/23 and runway 10/28, measuring 3660m X 45m and 1700m X 45m respectively. These lengths include 60m for a safety area called *Stop Way* (AIP, 2022).

The numbering of these runways is due to geographical orientation, following the instructions of international conventions. Both the main runway and the taxiways have the necessary signs and markings for night and day operations. All headlands have non-precision instrument approach procedures, but headlands 05/23 have, in addition to these systems, equipment called ILS, which provides pilots with a vertical glide slope via the airplane's instrument *display*.

The increase in precision due to this instrument allows for a reduction in the minimum visibility and ceiling conditions required for landings and take-offs. This system requires a set of approach lights called ALS (Approach Light System). These lights are installed at the headland and along the runway centerline.

In the case of Maputo International Airport, these can be seen by motorists driving along Av. 19 de Outubro, if they look at the supports set up inside the Mavalane Air Base compound. It is worth noting that these supports serve as perches for herons and seagulls.

**Figure 14: Image of the passenger embarkation/disembarkation bridges (domestic Terminal A),**

Source: Author, 2022

### 3.6. Other facilities

Other facilities worth mentioning are the Control Tower, the Fire Station and the fuel storage tanks. Kerosene for the turbine is taken to the aircraft via tankers and hydrants. Importantly, all the unpaved areas between the main runways and the taxiways, as well as the entire obstacle-free area along the main runways, are covered in grass.

Source: Author, 2022

### 3.7 Characterization of Air Traffic

The physical characteristics of Maputo International Airport make it possible to operate all types of traffic. It is possible to identify the operation of civil and military aircraft, scheduled air transport and general aviation, on domestic and international flights. It handles aircraft under instrument flight rules (IFR) and visual flight rules (VFR). Cargo aviation also has its place, as do helicopters. However, what predominates are scheduled air transport flights with medium to large aircraft.

### 3.8. Statistical data

The following statistics show that ARP/MPM is still an important link between the city of Maputo and other cities in the country, as well as with the main capitals of the world. These figures were obtained from the Maputo International Airport Operations Directorate and refer to 2019.

- Floating population: 900,000 people;
- Passenger movement/year: 2,174,691

- Air cargo movement/year: 10,747 tons

- Mail movement/year: 293.08 tons

- Landings and take-offs/day: 95 landings - 97 take-offs;

- Landings and take-offs/month: 21106 landings - 21089 take-offs;

- Landings and take-offs/year: 34,385 landings - 34,430 take-offs;

- Most used track: 05/23;

- Most commonly used procedure: IAL CHARLIE 3, header 05;

The peak times for landings and take-offs during the day are: between 18:00 and 20:00 (landing); and between 06:00 and 09:00 (take-off). The busiest months throughout the year are: January, March, October, November and December.

## 3.9. Socio-economic characterization of the sample

This subsection presents the respondents' social data, such as age, gender, education level, and their role within the company.

The present survey had a sample of 20 employees, of whom 73.3% were male and 26.7% female. With regard to the age of the respondents, it was found that they were aged between 25 and over 55, with the majority of respondents aged between 25 and 34 (60%), followed by respondents aged between 45 and 54 (20%) and 13.3% aged between 35 and 44 (6%).3% were aged between 35 and 44 and finally 6.7% were over 55, as shown in graph 1 below.

**Graph 1: Age of the population by gender.**

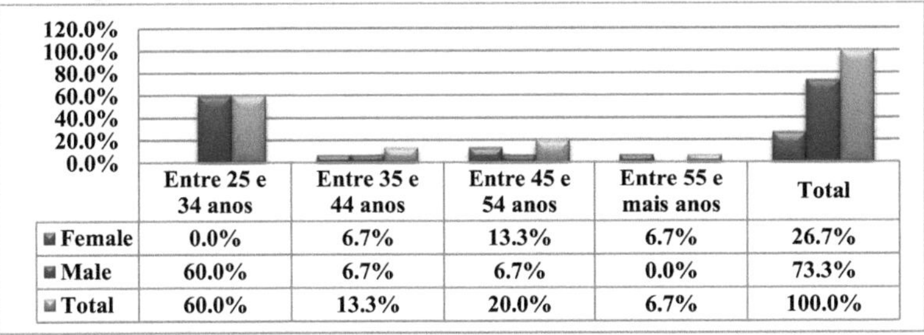

| | Entre 25 e 34 anos | Entre 35 e 44 anos | Entre 45 e 54 anos | Entre 55 e mais anos | Total |
|---|---|---|---|---|---|
| Female | 0.0% | 6.7% | 13.3% | 6.7% | 26.7% |
| Male | 60.0% | 6.7% | 6.7% | 0.0% | 73.3% |
| Total | 60.0% | 13.3% | 20.0% | 6.7% | 100.0% |

Source: Author through fieldwork, 2023

In order to analyze the respondents' educational qualifications, the variables gender and educational qualifications were cross-referenced in order to divide the respondents according to their level. From this cross-referencing, it was possible to see that most of the male respondents have a secondary education and a bachelor's degree, with around 26.7% respectively, while around 13.3% of the female respondents have a higher education degree, in this case a full bachelor's degree. Looking at the overall picture, around 40% of the respondents have a Bachelor's degree, followed by secondary education with 33.3% and finally a Master's degree and other training with 13.3% respectively.

In research related to risk, education plays a very important role, because in order to better understand the dynamics, as well as make decisions about certain phenomena that may occur in and around airports, it is necessary for the respondent to have a considerable level of education and, looking at the data presented here, it is possible to discern that the respondents have an educational background that allows them to make decisions in line with what is required.

**Graph 2: Qualifications of the population by gender.**

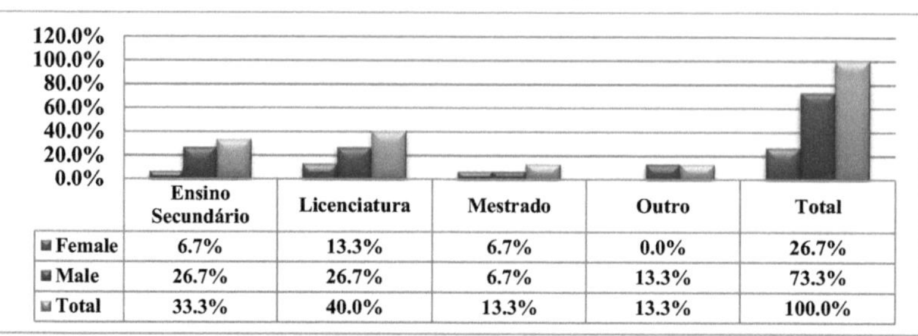

| | Ensino Secundário | Licenciatura | Mestrado | Outro | Total |
|---|---|---|---|---|---|
| ▪ Female | 6.7% | 13.3% | 6.7% | 0.0% | 26.7% |
| ▪ Male | 26.7% | 26.7% | 6.7% | 13.3% | 73.3% |
| ▪ Total | 33.3% | 40.0% | 13.3% | 13.3% | 100.0% |

Source: Author through fieldwork, 2023

When asked about the position they work in, variables were defined in which respondents were asked to state in which part or area of the airport they work, with the following alternatives *"Inside the airport building, outside the airport building and both parts",* where the majority of respondents showed that they only work inside the airport, with around 53.3%, and in turn around 46.7% of respondents showed that they work both inside the airport and outside it. With this data, it can be concluded that most of the respondents experience the accidents that occur at the airport, which may mean that they have the autonomy to respond to the different collisions between birds and aircraft at Maputo International Airport.

**Graph 3: Area where you carry out your activities**

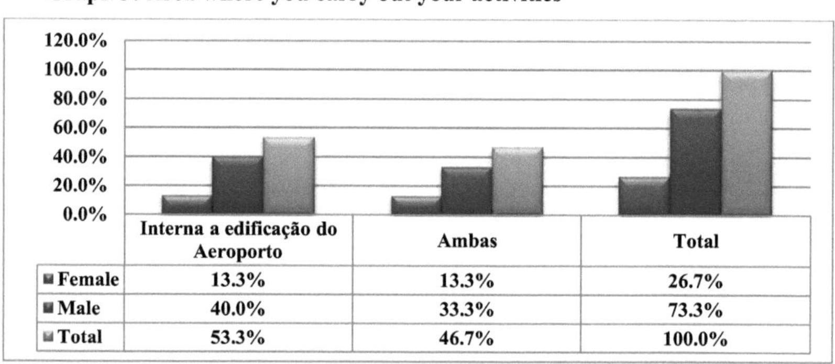

| | Interna a edificação do Aeroporto | Ambas | Total |
|---|---|---|---|
| ▪ Female | 13.3% | 13.3% | 26.7% |
| ▪ Male | 40.0% | 33.3% | 73.3% |
| ▪ Total | 53.3% | 46.7% | 100.0% |

Source: Author through fieldwork, 2023

The data presented in Graph 4 shows the cross-reference between the function of the survey respondents and the length of time they have been working at the airport and in their careers. From the data collected in the field, it was possible to see that the respondents have worked at the airport for between 10 years and more than 31 years and that they are pilots, air traffic controllers, airport managers, airport operators and other related areas.

With regard to the length of time they have worked at the airport, it was possible to conclude that most of the respondents have worked at the airport for up to 10 years (46.7%), of whom 20% are air traffic controllers and 6.7% work in the other areas mentioned above, followed by respondents who have worked at the airport for between 21 and 30 years (26.7%).7%, of whom 6.7% work as air traffic controllers, airport managers, airport operators and other related areas, and finally 13.3% work at the airport for between 11 and 20 years and more than 31 years, some of whom are air traffic controllers and others airport operators.

**Graph 4: Length of time working at the airport.**

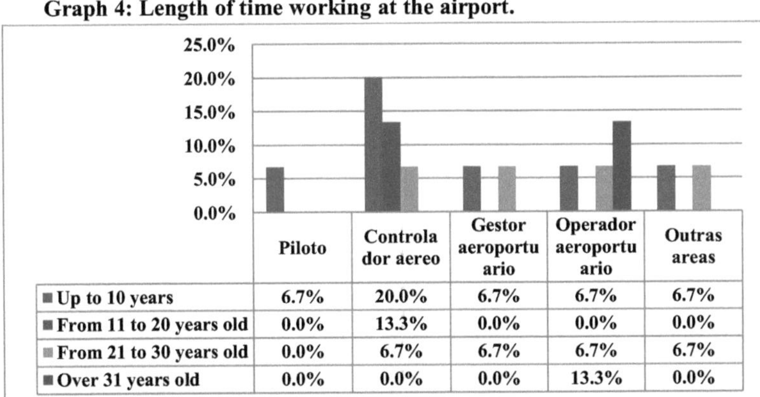

| | Piloto | Controlador aereo | Gestor aeroportuario | Operador aeroportuario | Outras areas |
|---|---|---|---|---|---|
| Up to 10 years | 6.7% | 20.0% | 6.7% | 6.7% | 6.7% |
| From 11 to 20 years old | 0.0% | 13.3% | 0.0% | 0.0% | 0.0% |
| From 21 to 30 years old | 0.0% | 6.7% | 6.7% | 6.7% | 6.7% |
| Over 31 years old | 0.0% | 0.0% | 0.0% | 13.3% | 0.0% |

Source: Author through fieldwork, 2023

**3.10. Wildlife risk at Maputo International Airport in the view of the respondents.**

Regarding the sighting of animals in the property area, the respondents were unanimous in stating that they had already sighted animals in the airport area, and this data was cross-referenced with the function and in this cross-referencing it was possible to see that the pilots were the ones who most sighted animals in the property area of Maputo International Airport with about 33.3%.3%, followed by Airport Operators with 20%, while air traffic controllers spotted animals with around 13.3% and finally we have the Airport Managers as those who had the fewest sightings of animals inside the airport with a total of 6.7%, it should be noted that other respondents from other areas stated that they had never seen animals on the runway.

**Graph 5: Animal sightings in the property area or on the Maputo Airport runway**

| | Gestor aeroportuário | Piloto | Controlador aéreo | Operador aeroportuário | Outras áreas |
|---|---|---|---|---|---|
| Yes | 6.7% | 33.3% | 13.3% | 20.0% | 0.0% |
| No | 6.6% | 6.7% | 0.0% | 6.7% | 6.7% |

Source: Author through fieldwork, 2023

Regarding the participation of the airport community in lectures and training on the risk of aircraft collision with fauna, fauna conservation and on procedures to prevent animals from approaching the airport's operational area, the results showed that airport operators are the ones who have participated most in lectures at the airport on the treatment that should be given to waste in the last 5 years with around 13.3%, followed by pilots and staff from other areas with around 6.7% respectively, giving a total of 26.7% of employees who have benefited from a lecture, where the focus is on the risk of a collision between aircraft and wildlife (Graph 6). In general, this data allows us to infer that most of the employees at Maputo International Airport are not included in the ongoing training plans, since around 73.3% of those surveyed said that in the last 5 years they had not attended any lectures on the treatment of waste and fauna within the airport.

**Graph 6: Attendance at lectures at the airport on how waste should be treated in the last 5 years.**

| | Piloto | Controlador aereo | Gestor aeroportuario | Operador aeroportuario | Outras areas | Total |
|---|---|---|---|---|---|---|
| Yes | 6.7% | 0.0% | 0.0% | 13.3% | 6.7% | 26.7% |
| No | 0.0% | 40.0% | 13.3% | 13.3% | 6.7% | 73.3% |
| Total | 6.7% | 40.0% | 13.3% | 26.7% | 13.3% | 100.0% |

Source: Author through fieldwork, 2023

### 3.11. Level of Wildlife Risk at Maputo International Airport

Land use change near airports can affect animals, subsequently contributing to the risk of collisions (CONKLING et al., 2018). In relation to the risk related to fauna at the airport, participants were asked to classify the level of risk of fauna in the study area, to which about 70% of the participants classified the level of risk of fauna present at Maputo International Airport as being High Risk, in turn 26.73% classified the risk as being the risk is medium, for 3.3% of the participants of this research classified the risk is low, as illustrated in the graph below.

**Graph 7: Fauna Risk Level at the Airport**

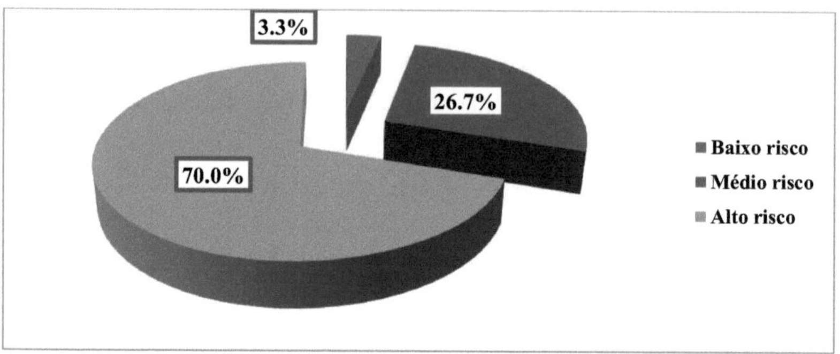


Source: Author through fieldwork, 2023

For the majority of participants, the risk of fauna present at AIM varies between medium and high. For those who frequent aircraft movement areas, the constant presence of birds in these areas is common.

Regarding the frequency of animal sightings at the airfield, the survey shows that respondents see animals at the airfield frequently and rarely with 40% respectively and others always see animals at the airfield with 20%, it should be noted that whether or not wildlife is seen at the airport depends on the area where each employee works within the airport.

**Graph 8: Frequency of animal sightings at the airfield.**

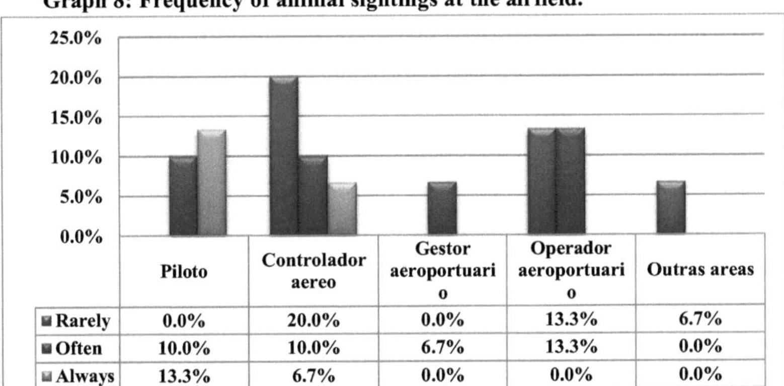

|  | Piloto | Controlador aereo | Gestor aeroportuario | Operador aeroportuario | Outras areas |
|---|---|---|---|---|---|
| Rarely | 0.0% | 20.0% | 0.0% | 13.3% | 6.7% |
| Often | 10.0% | 10.0% | 6.7% | 13.3% | 0.0% |
| Always | 13.3% | 6.7% | 0.0% | 0.0% | 0.0% |

Source: Author through fieldwork, 2023

Avian danger is a major factor in the flight safety system at airports and has generally not been sufficiently taken into account by the municipal authorities, who are primarily responsible for controlling urban land use and commercial activities around the airport, and who go unpunished for their actions. The lack of action by urban authorities has led to the appearance of degraded environmental and sanitary conditions in the areas surrounding airports, contributing significantly to worsening the problem by increasing the incidence of bird attractants due to the supply of organic material.

With regard to the influence of environmental conditions on the occurrence of aviation accidents caused by birds, 60.1% of respondents said that yes, conditions do influence the occurrence of collisions between birds and aircraft, because when environmental conditions are degraded they can lead to the appearance of hotspots for birds and other animals, and around 39.9% said that environmental conditions do not influence the occurrence of collisions between birds and aircraft.

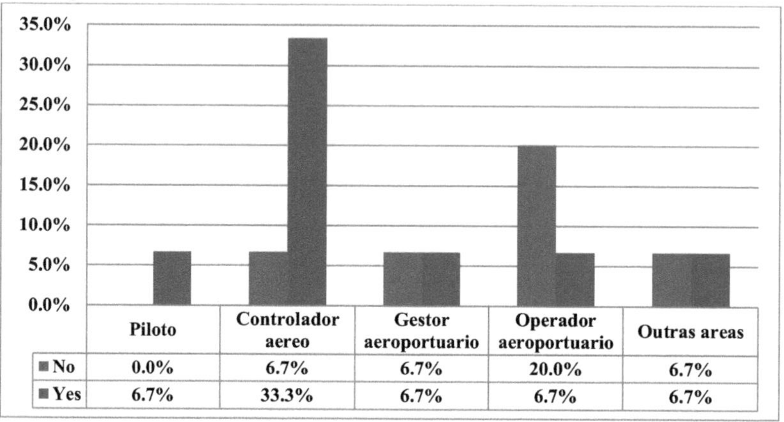

| | Piloto | Controlador aereo | Gestor aeroportuario | Operador aeroportuario | Outras areas |
|---|---|---|---|---|---|
| ■ No | 0.0% | 6.7% | 6.7% | 20.0% | 6.7% |
| ■ Yes | 6.7% | 33.3% | 6.7% | 6.7% | 6.7% |

Source: Author through fieldwork, 2023

In order to avoid accidents related to fauna and especially birds, it is necessary that the measures taken in the process of managing the risk of avian fauna and other types of fauna are efficient or effective, since the inefficiency of these measures can compromise the entire existing management process within an airport and in particular the AIM in this regard, We tried to find out from the respondents whether the fauna management measures in the AIM are efficient or not, and it was possible to see that most of the respondents (Pilot, Air traffic controller, Airport manager, Airport operator, Other areas) said that the fauna management or control measures in the AIM are not very efficient, with around 60%, while another group said that the measures are not efficient, with 33.3%, and finally only 6.7%.3% and finally only 6.7% of those surveyed said that the measures are efficient. It should be noted that of the group of people surveyed, the controllers are the ones who stood out with 27.7% saying that the measures are inefficient.

**Graph 10: The fauna risk management measures adopted at the aerodrome are efficient.**

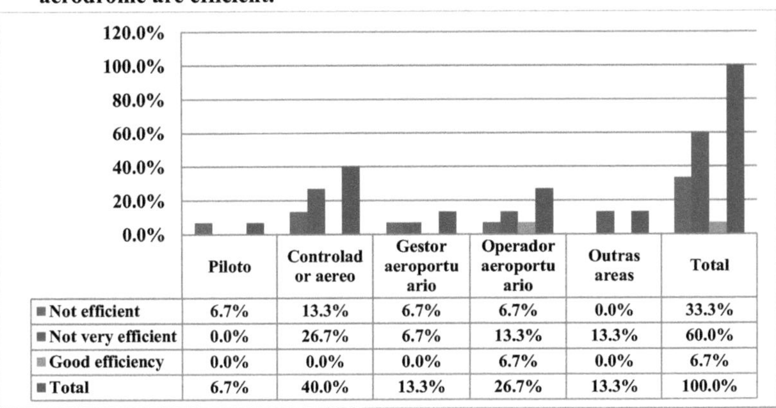

| | Piloto | Controlador aereo | Gestor aeroportuario | Operador aeroportuario | Outras areas | Total |
|---|---|---|---|---|---|---|
| ■ Not efficient | 6.7% | 13.3% | 6.7% | 6.7% | 0.0% | 33.3% |
| ■ Not very efficient | 0.0% | 26.7% | 6.7% | 13.3% | 13.3% | 60.0% |
| ■ Good efficiency | 0.0% | 0.0% | 0.0% | 6.7% | 0.0% | 6.7% |
| ■ Total | 6.7% | 40.0% | 13.3% | 26.7% | 13.3% | 100.0% |

Source: Author through fieldwork, 2023.

When asked what action they had taken when they spotted an animal in any of the AIM areas, the majority of respondents said that they had called the fire department and the environmental agencies, with around 25% and 20% respectively. Another group said that they had notified some airport staff after spotting animals at the airport. These choices of attitudes or measures can be explained by the responsibilities inherent in their duties to keep the operational area free of animals, to avoid the risk of collisions with aircraft. However, depending on their area of operation, concessionaires and contractors also have responsibilities in terms of aviation safety.

**Graph 11: Attitudes taken when spotting an animal**

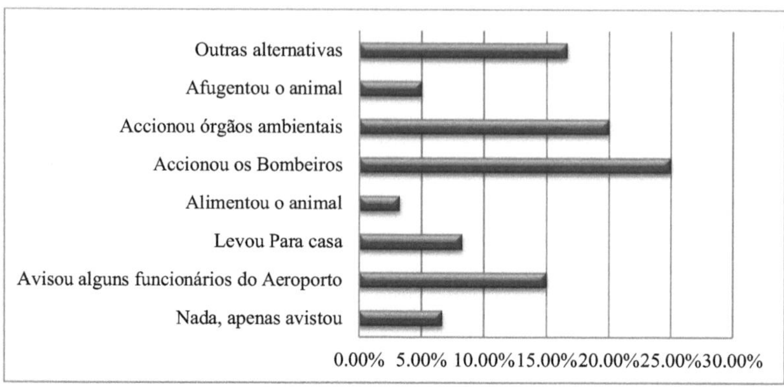


Source: Author through fieldwork, 2023

This result infers that many people have adopted measures to mobilize the removal of animals present in the operational areas, which may be due to the commitment and attributions of each category within the airport. This area, although restricted, is visible to everyone heading airside due to the large glass windows separating the passenger terminal and the operational airside, so anyone who sees an animal in the movement area could report it to airport staff in order to take immediate action.

### 3.12. Factors influencing the risk of collision between aircraft and birds

The occurrence of accidents related to collisions between avian fauna and aircraft is caused by various factors. In addition to activities of a dangerous nature, other factors cause an ecological imbalance that is a determining factor in increasing the avian risk at the airport, even if indirectly, such as, the areas used to dispose of solid urban waste, usually in the open, located along Avenida Gago Coutinho right at the final approach to runway 05 and the fence on Rua da Beira, the Vulcano and Mavalane markets with all their pollution from commercial activities. The presence of residential areas arranged in a disorderly fashion, next to the fence, with precarious or no basic sanitation infrastructure, already contributes to this picture.

This survey sought to find out from the respondents what the main conditions were that influenced the proliferation of birds and consequently the risk of a collision between birds and aircraft at AIM. Most of the respondents pointed to the existence of the Hulene dump and the concentration of solid waste, markets and stalls around the airport as the main factor, with 26.7%, because according to them, these are places where birds and other animals look for food. They then pointed to the existence of a snack bar inside the airport and the cultivation or planting area around the airport with 17.3% and 18.7% respectively.

**Graph 12: Main factors influencing the risk of collision between birds and aircraft.**

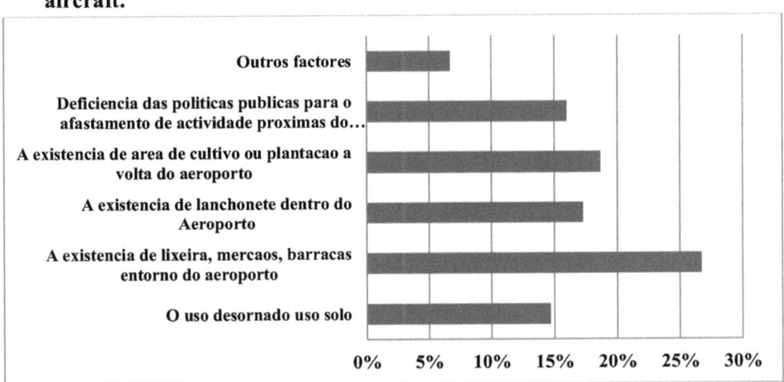


Source: Author through fieldwork, 2023

### 3.13. Fauna attraction factors and hotspots.

According to Ferreira (2011), the main factor for the presence of birds on airfields or in nearby areas is the search for food, water, shelter and rest, and looking at the physical characteristics of airfields, these are favorable for the permanence of various species of birds.

Bearing in mind that the environment is a factor that influences the behavior of fauna species that remain in search of food, water and shelter, this item seeks to associate the existence of birds in the ASA and their possible attraction points so that measures can be defined to mitigate or control the risk they pose.

Map 4 shows the Airport Security Area of Maputo International Airport, an area that covers Maputo City, Matola, the Marracuene District and Matutuine. Within this area, several potential wildlife attractions have been identified, most of which are within a 5 km radius, as Map 4 illustrates. Among the various potential wildlife or bird attractions, the following stand out: Hulene garbage dump, residential neighborhoods near the airport fence, zoo, solid waste disposal areas, municipal rainwater collector, incinerator, wastewater treatment plant, municipal slaughterhouse, Vulcano market, Malhazine granary, Maputo Port cereal terminal, Malauzi River, Joaquim Chissano Avenue drainage ditch and agricultural areas.

It should be noted that in addition to these there are other potential attractions along the 20 km length of the ASA, but here we will focus on identifying the potential attractions that are within the 5 km sighting radius, as illustrated by the map below.

Map 2: Location of the main wildlife hotspots in the ASA of Maputo International Airport

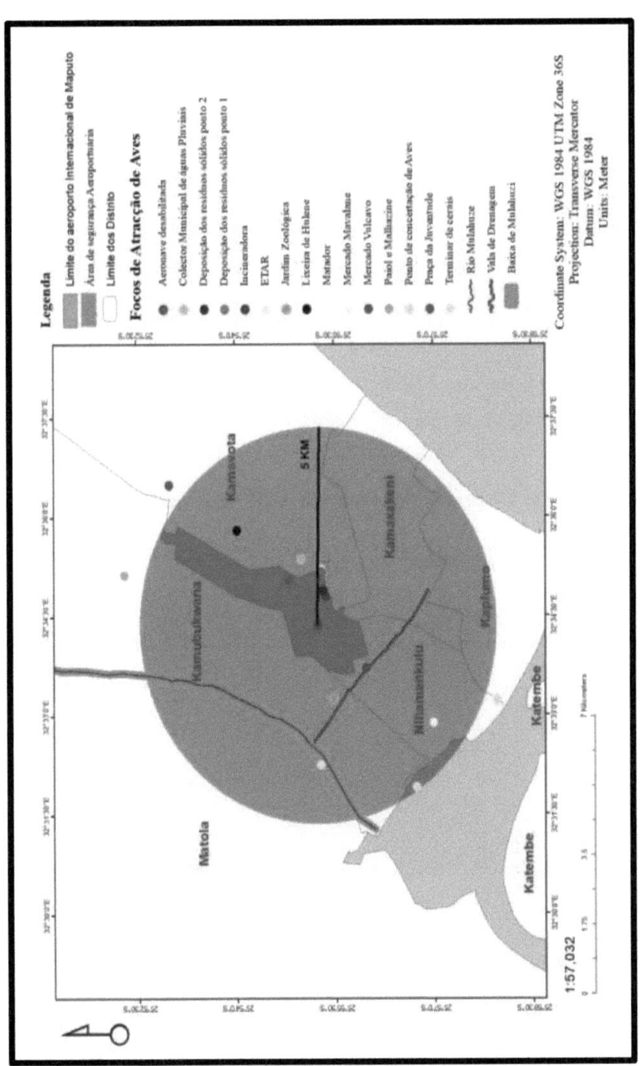



Source: Author through fieldwork, 2023

The data presented here corroborates the research carried out by CLEARY and DOLBEER (2005), who state that the main factors attracting birds are the commercial activities typical of the airport environment, such as restaurants and commissaries, when carried out without concern, the existence of areas for the disposal of organic waste generated, as well as the existence of areas where birds can find food and water to drink, among other factors contribute to the increase of this waste in airport areas, with a consequent increase in the population of birds that are attracted.

The images below are examples of some of the bird hotspots, where you can see that just under 5 km from Maputo International Airport, there is a WWTP 2,700 meters from headland 10, Lhanguene Cemetery 2,660 meters from headland 05, the zoo 870 meters from headland 10, Hulene Dump 595 meters from headland 23, and the Merchandise Station 1,150 meters from headland 28.

**Figure 16: Distances between the AIM and some potential bird hotspots.**

Source: Google Earth, 2023.

Figure 17: Distance from Hulene dump to Headland 2

Source: Author - Google Earth, 2023.

**Figure 18: Solid waste disposal areas inside the airport.**

Source: Author, 2023.

**Figure 19: Birds at the Hulene rubbish dump**

**Source:** DW- Marta Barroso

### 3.14.Management of AIM airport space.

With regard to accidents involving collisions between birds and aircraft, the ICAO recognizes, however, that the airport operator has limited efficiency, since birds easily pass through the approach and take-off paths, where other authorities must collaborate to prevent avian risk. This is particularly true of the municipal authorities that manage solid waste and control economic activities.

**Figure 20: Bird flight height/priority areas to avoid collisions**

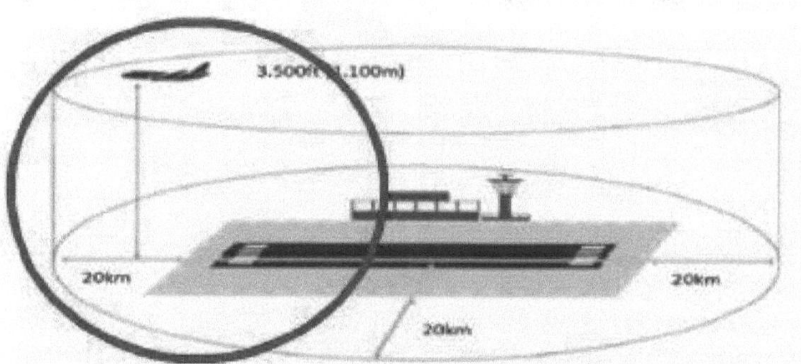


Source:https://pt.linkedin.com/pulse/risco-de-fauna-na-avia%C3%A7%C3%A3o-birdstrike-eduardo-ortiz-pereira, 22-09-2023.

It is clear that everyone wants to reach their destination safely when making any kind of journey, exercising their right to come and go freely. However, the internal question would be what is the responsibility of the airport operator in this context, and the answer is that it has a clear responsibility. After all, the majority of collisions occur in or near airport premises, a fact corroborated internationally by the various documents issued in this regard. This highlights the responsibility for developing, implementing and publicly displaying an efficient bird and wildlife collision control program, appropriate to the size and complexity of the airport, taking into account the identification of avian risk and its assessment (ICAO, 2011:3-1).

The appropriate authority shall act to eliminate or to prevent the establishment of solid waste disposal sites or any other source that may attract animals to the aerodrome or its vicinity, unless appropriate assessment indicates that such sites are unlikely to create conditions that lead to wildlife problems. Where elimination of existing sites is not possible, the appropriate authority shall ensure that the risks caused to aviation by such sites have been assessed and reduced to the lowest practicable risk condition (ICAO, 2009: 9-10).

After stating that collisions between birds and aircraft occur on the AIM, respondents were asked about the effects that these collisions cause, where respondents pointed to two consequences or effects of these collisions which were aborted take-off with 53.3% and emergency landing with 46.7%, as illustrated in Table 2 below.

**Table 2:** Effects of collisions between birds and aircraft on AIM.

|  |  | N | Percentage |
|---|---|---|---|
|  | Aborted take-off | 16 | 53.3% |
|  | Emergency landing | 14 | 46.7% |
| Total |  | 30 | 100.0% |

Source: Author through fieldwork, 2023

### 3.15. Strategies and Techniques for Bird Control at Airports or in the Event of a Collision.

Bird control at airports is a very dynamic field. New products and technologies are being developed and environmental laws are being improved. Therefore, these strategies and techniques are not the final solution to the problem, but rather a tool for an appropriate bird management plan at airports. With regard to the strategies used or taken to control birds within the airport, the survey showed that, according to the respondents, in the event of an accident or collision between birds and aircraft, the AIM uses two mechanisms and strategies to control the situation, which are changing the flight schedule or using techniques or methods to scare birds away using Sound Deterrent

Artifices, with 33.3% respectively. Respondents also pointed out that the AIM also uses visual deterrents to scare animals away from the aerodrome, such as flags and banners as well as mirrors, models of birds such as hawks and other birds of prey with 26.7% among other options, as illustrated in Table 3 below.

**Table 3:** Animal control measures in the event of accidents or sightings at the aerodrome.

|  |  | N | Percentage |
|---|---|---|---|
|  | Changing flight schedules | 15 | 33.3% |
|  | Changing bird habits and habitats | 3 | 6.7% |
|  | Deceptive devices | 15 | 33.3% |
|  | Visual deterrents | 12 | 26.7% |
| Total |  | 45 | 100.0% |

Source: Author through fieldwork, 2023

### 3.16. Bird Hotspots and Associated Species in the Security Area (ASA) of Maputo International Airport.

ICAO Doc 9137 Part 3 defines the ASA as the extent covered by a pre-established radius, depending on the type of operation of the airport and drawn from the "geometric center of the aerodrome. This varies from a radius of 20 km for aerodromes operating under instrument flight rules (IFR) to a radius of 13 km for other aerodromes.

This area, according to ICAO recommendations, should have a radius of approximately 20km, and according to the same source and other competent bodies, they recommend the creation of land use restrictions, bearing in mind that the negative effects of noise and irregular construction can lead to problems in air operations. According to ICAO (2001), landowners are prohibited from building, planting, elevating or carrying out any work on the property that would encroach on the space delimited by the laws in force. They may carry out any activity within the zone if they have prior authorization from the government.

Map 3: Delimitation of the Airport Security Area at Maputo International
Airport.

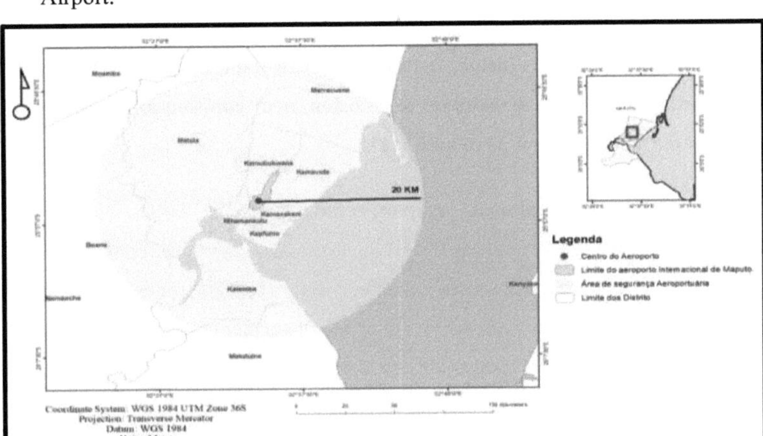


Source: Author through fieldwork, 2023

According to the MGRF, a major problem is that most collisions with
wildlife occur in the Airport Security Area (ASA), indicating a priority
environment for wildlife risk management actions, according to Law
12.725/2012 (CENIPA, 2017).

On average, 1 to 2% of the ASA is in the aerodrome asset area. The
consequence is that there are two different parties responsible for wildlife risk
management in this area. The public authority with the area within the ASA
and the aerodrome operator are responsible for urban development policy, the
aim of which is to order the full development of social functions and guarantee
the well-being of its inhabitants.

It is up to the municipality, within the scope of its urban policies, in the
role of land use and occupation planning, to prevent the installation of activities
in the ASA that could become attractive hotspots for fauna and, consequently,

cause risks to air operations (DEMANDS AND GUIDELINES ON CIVIL AVIATION, 2016).

### 3.17. Risk of collision between aircraft and birds at Maputo International Airport: assessment of collision, near collision and sighting reports on AIM between 2016 and 2021.

Maputo International Airport, Mozambique's largest airport infrastructure, plays a key role as one of the main gateways to the country and an important national and international transportation hub. Its strategic location offers easy access to the main cities and economic centers of the country and the region, with regular connections to destinations in Africa, Europe and South America. Since the airport was built in 1960 and has been in operation, wildlife issues have been documented and recorded on its own spreadsheets, as required by ICAO, where monthly reports are produced with statistics on the different species of fauna most frequently sighted on the airport grounds, both during routine patrols and sightings on the airport's runways and maneuvering areas. It should be noted that of the data provided by the AIM, we only have data on collisions between birds and aircraft, as shown in the table in Annex 1.

From 2016 to 2021, according to data that the research had access to, Maputo International Airport recorded a total of 34 collisions with different species of birds, with the Cape Caracara *in Portuguese with the scientific name "Burhinus capensis" or* Spotted Thick-knee *in English and the Persian Bee-eater* in Portuguese with the scientific name *"Merops persicus Sandpiper" or* Blue-cheeked Beeeater, which are both regional and planetary migratory species according to Decree-Law no. 51/2021 of 19 July, approving the Regulations for the Protection, Conservation and Use of Birdlife.° 51/2021 de 19 de Julho, which approves the Regulation for the Protection, Conservation and Sustainable Use of Avifauna, were the species that between 2016 and

2021, according to data provided by the AIM, had the highest number of collisions between birds and aircraft, where these two species of birds together totaled around 18 collisions, which represents around 52.9%, with the species "Burhinus capensis" colliding with aircraft more than 9 times and the bird of the *"Merops persicus"* family *colliding with aircraft a total of 5 times.*

In the period under review, 3 collisions between birds and aircraft were recorded in January and September 2016, 6 collisions were recorded in January, September, November and December 2017, with December having the highest number of collisions, February, March, April and December 2018, AIM recorded a total of 13 collisions and February had the most collisions with a total of 7, in 2019, around 11 collisions between birds and aircraft were recorded in the months of February, August, September, November and December, with November recording the highest number of collisions with 5. Finally, in 2021, only one collision was recorded in the month of June, according to data available on the AIM and consulted in 2021, as shown in the table in Annex 1.

The data related to collisions between birds and aircraft at AIM and operations recorded at the same airport allowed us to calculate the collision rate between 2016 and 2021 at Maputo International Airport, using Equation I, as shown in Table 4.

**Table 4:** Crash rate per year at Maputo International Airport

| Year | Number Of Operations | Number of collisions | Collision rate |
|---|---|---|---|
| **2016** | 21.545 | 3 | 0.139 |
| **2017** | 18.858 | 6 | 0.318 |
| **2018** | 20.116 | 13 | 0.646 |
| **2019** | 21.106 | 11 | 0.521 |
| **2020** | 11.088 | 5 | 0.450 |
| **2021** | 14.378 | 1 | 0.069 |

Source: AIM data, 2023.

The results obtained by applying the formula for evaluating the bird-aircraft collision rate reveal an oscillation between the number of operations and the number of collisions, as 2016 had around 21,545 operations and only 3 collisions were reported, while 2018 had around 20.116 airport operations and 13 collisions were reported, 10 more than in 2016 and with fewer operations, which shows that the number of collisions is not related to the number of operations at AIM, there is another factor that causes collisions to be frequent at Maputo International Airport, these reasons may be related to the environment surrounding AIM, such as growing urbanization and disordered land use, with the emergence of inadequate waste deposits and other focuses of attraction for fauna.

In this way, the application of this index corroborates the assertion that the increase in the number of collision records may be a positive sign of greater involvement and commitment on the part of those involved in civil aviation safety with wildlife risk management, especially air operators, whose pilots are the first to become aware of the collision event during the flight.

### 3.18. Birdlife Risk Assessment

Taking into account the methodologies presented for assessing the risk of collisions between aircraft and birds at Maputo International Airport, as well as for identifying problem species, according to Graph 13, it was possible to see that the AIM in terms of the risk of collisions between birds and aircraft has three levels of risk that vary between very high and medium risk, with very high risk having the greatest representation with around 40%, followed by high risk with 33.3% and finally medium risk with around 26.7%.

**Graph 13: Risk assessment of fauna associated with AIM, using Methodology I.**

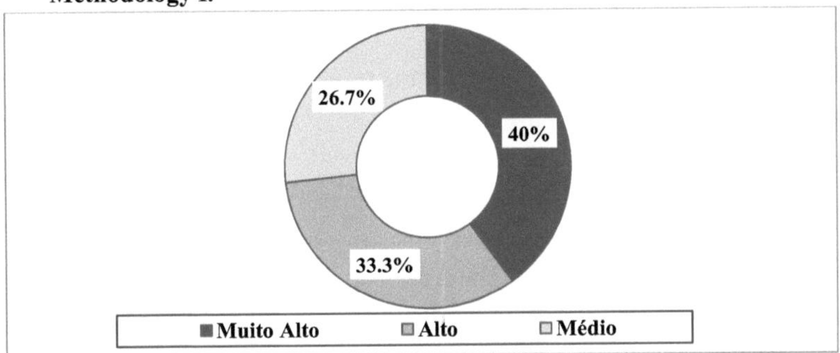


Source: Author (2023) using data from 2016 to 2022 provided by AIM.

One of the factors influencing the analysis of avian risks at aerodromes is related to the environmental factors of the development as well as the environmental conditions of the surrounding area. According to the data collected in this study, Maputo International Airport has an ASA area with many bird attraction points, including a wastewater treatment plant and a rubbish dump, which are located in the aerodrome's final approach area. In this sense, the high levels of risk presented by the AIM in this study are related to the environmental conditions of the surrounding area.

With regard to the problem species at the AIM, according to the risk assessment matrix in Annex 1, the problem species that present a very high risk, i.e. those most likely to cause an accident at the airport, are: *Herons, Caspian tern / great egret, Ardea melanocephala / black-headed egret, Burhinis capensis / Cape scarab, Stork Accipiter minullus / little hawk.*

The species at high risk were: *Swallow, Harrier Circus/ Hawks, Charadrius tricoloris/ Three-collared Sandpiper, Thickbilled lark, Vidua regia/ Queen whydah,* and in turn the *Owl, Merops persicus/ Persian Bee-*

*eater, Polyborides typus/, Terek sandpiper/ Sandpiper* in terms of probability of causing an accident between birds and aircraft presents a medium risk.

### 3.18.1 Analysis of the results of the fauna risk assessment at Maputo International Airport.

In order to assess the risk of collisions between aircraft and birds at AIM, two methodologies were applied, the main aim of which was to assess the risk at the aerodrome, with a view to defining control actions to be designed and implemented to reduce or eliminate the risk of collisions between birds and aircraft at the aerodrome.

Methodology I, which focused on calculating the collision rate between aircraft and birds at the AIM aerodrome between 2016 and 2021, taking into account the data provided, it was possible to see that the variation in the collision rate is not directly related to the number of boarding or disembarking operations, as there were years in which a greater number of operations were recorded but the number of collisions is reduced, as is the case of the year 2016 in which a record of 21.545 operations and only 3 collisions were recorded, while in 2018 there were around 20,116 operations and 13 collisions. This increase in the collision rate is related to the environmental conditions on the airport site and in its surroundings, which are conducive to the creation of mooring points for an even greater number of birds in the ASA area, which can increase the frequency of collisions between birds and aircraft, thus increasing the collision rate even though there has been a reduction in operations at the airport.

The herons Caspian tern, Ardea melanocephala, Burhinis capensis, Cape curlew and Accipiter minullus have all been reported to have collided at the airfield, with one collision having an effect on the flight. The Burhinus capensis" has collided with aircraft more than 9 times and the bird of the

"Merops persicus" family has collided with aircraft a total of 5 times between 2016 and 2021, as it presents a risk to operations due to its constant presence in aircraft maneuvering areas, including during aircraft landing, parking and take-off phases. The Heron, which forms flocks near the runway, has also been recorded as a problem species. All of these species are related to hotspots present at the airport or in its ASA.

The results given by methodologies I and II were able to highlight the greater risk of these species at Maputo International Airport, as they were able to accurately assess the likelihood and impact that could be caused by a collision with each species.

Fauna monitoring activities must be carried out with standardization in order to achieve a level of understanding of animal dynamics, and also have qualified personnel to understand bird behaviour. Hence the need to create regulations establishing the Control of Fauna in the Airport Security Area (ASA), the proposal for which is presented in Annex 2, so that the attractions present around them can be minimized.

**Figure 21: Flight level limit in the ASA**

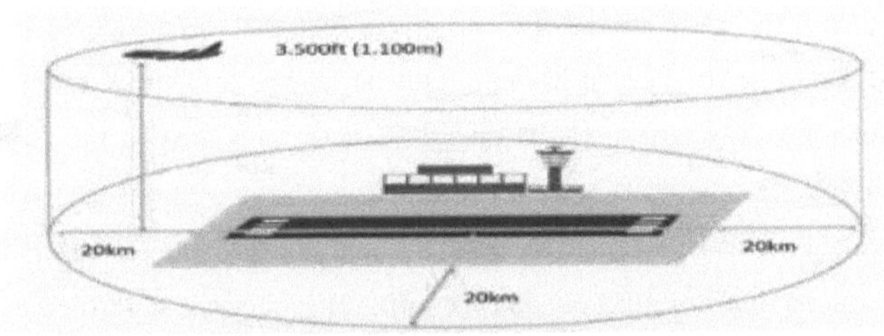


Source: Fly Safe, published on 05/11/2020

In addition, the proposed regulation defines rules to reduce accidents, which activities can attract animals, responsibilities, penalties, among others.

# CHAPTER 4: CONCLUSIONS

This research sought to analyze the influence of environmental conditions around airfields and their possible effects on the attraction of avifauna to these sites and within the infrastructure of Maputo International Airport in the years 2016 to 2021. Exploratory research with a qualitative approach was used to achieve the objective. Bibliographic and documentary research was used to collect data and information.

This survey was carried out with a sample of 20 employees made up of pilots, air traffic controllers, airport managers, airport operators and others who work both inside and outside the airport, all of whom have been working at Maputo International Airport for more than 10 years. With regard to animal sightings and specifically birds, it was possible to see that air traffic controllers, pilots and airport operators are the ones who have seen birds, with pilots seeing them more often.

According to the results of the research into the level of risk related to fauna at Maputo International Airport, it is classified as high, as it is believed that inside as well as around the airport, conditions are created for attracting birds, since within a radius of 20 km, which is the radius defined for the Airport Security Area, there are several infrastructures that can be considered to be hotspots, including the Port of Maputo, Hulene Dump, Infulene Wastewater Treatment Plant, Lhanguene Cemetery, residences built next to the airport fence, among other hotspots, which contributes to bird collisions at Maputo International Airport between 2016 and 2021, and reinforces the likelihood of the collision having occurred outside the airport perimeter and verified after the aircraft has landed, where the greatest focus is needed to avoid the presence of fauna, as it is considered to be a sighting zone.

According to the results of the survey, there is no doubt that collision reporting is a way of increasing aviation safety at airports and in areas close to the airport. It is a fact that when something out of the ordinary (accident/incident) happens during a flight, everything needs to be reported as a way of creating an internal safety culture, helping to prevent accidents and improving internal processes.

In relation to the risk assessment with the application of assessment methodology I and II, the results obtained allowed us to conclude that the collision rates between birds and aircraft tend to increase even with the decrease in airport operations within Maputo International Airport, which shows that there is an increase in the frequency of bird visits within the ASA, as a consequence of the growing increase in bird hotspots within the airport security area.

Finally, it was possible to see that in terms of risk level looking at the problem species, Maputo International Airport presents a very high risk, and the problem species identified in this research are the Herons, Caspian tern, Ardea Melanocephala / Black-headed Heron, Burhinis capensis / Cape Curlew, Stork Accipiter minullus / Lesser Sparrowhawk, all of which have a history of colliding with aircraft and are related to the hotspots in the ASA and inside the airport.

Interventions are therefore needed in the airport environment and its surroundings                    in                    order                    to environment, in order to control the main hotspots identified, especially those whose operation is not in compliance with current legislation, and thus reduce the risk associated with the problem species identified. The entities responsible for the airport must take special care and attention during the hottest time of the year - summer, with clear skies. Data shows that this is the time when collisions are most likely to occur, so care must be intensified at this time of year.

It was found that Maputo International Airport has guiding instruments for the management and control of fauna hazards, namely: the Fauna Risk Management Plan (PGRF), the Airport Environmental Management Plan and the Aerodrome Manual, which consists of an analysis of data collected by a biologist specialized in fauna control, in order to ensure correct management and thus prevent *Bird Strikes* from occurring. However, the team that makes up the Environmental Management Unit does not include a Biologist or Ecologist whose job it would be to identify the type of bird and define the problem species

It is suggested that the Civil Aviation Authority and Maputo International Airport Administration, which has responsibility for preserving the operational conditions of the airport and maintaining the operational safety of aircraft, create awareness-raising tasks for the population located in the areas around the aerodrome, such as:

- Production of pamphlets alerting people to the risk of collisions between aircraft and birds, proposing simple preventative measures that can help reduce the number of species in the vicinity of the airport, such as, for example, when visiting airport areas, avoiding disposing of any kind of garbage or food waste on the ground;

- To visit schools and, through lectures, present the topic to the students, thereby disseminating knowledge and contributing to cultural development;

- To train the professionals who make up the Environmental Management Unit working at airports to be able to recognize the species involved in the collision, as identification is important for the development of research into each species, facilitating the creation of new, increasingly efficient control techniques;

- Investing in equipment such as binoculars and cameras to record images of the birds seen at airfields, so that specialized people such as biologists and ecologists can make a more detailed identification;

- Drawing up projects for community visits to the airport site, i.e. showing in practice how important avian danger is for the safety of air navigation and everyone (people and infrastructure) around the airport;

This is why it is so important to build this relationship between government bodies, municipalities, the community and airport staff, creating an environment where people are aware of the seriousness of the problem and working together to prevent it as the most effective way of reducing collision rates.

It can therefore be concluded that the risk associated with avian danger can and must be managed and combated by everyone, since a collision between aircraft and bird can have catastrophic consequences for the operator, users, residents of the area around the airport and the environment.

This highlights the need to work on the day-to-day life of the area around the airport, showing how important social influence is for aviation and the creation of risk management programs with the participation of the community.

As a suggestion, further research could be carried out at the country's main airports certified by ICAO to receive international traffic, in which each airport manager, together with the local authorities, could analyze the risks and present them to the municipal manager, in order to implement appropriate prevention and control measures for each geographical area.

In the course of the research, some limitations were encountered, which had to do with the difficulty in obtaining more consistent samples on the occurrence of collisions between aircraft and birds at the airport, since most of the events recorded between 2016-2021 were not reported, those reported did not present complete data, and the type of bird involved, size and location (position) of the collision were not identified. This condition is fundamental for assessing risk, controlling avifauna and establishing appropriate control measures, as well as for identifying the species that are a problem for air operations at Maputo International Airport.

The false perception that nothing can be done to control wildlife risk in much of the aviation sector generates little investment in training personnel to carry out fundamental management actions in various organizations directly involved with this problem.

The lack of openness on the part of the institutions to provide data for research and the culture of not making public the activities carried out in the aviation sector were barriers to data collection.

Lack of effective implementation of the procedures described in the fauna management instruments available at the airport, verified by the existence of tall grass, bushes whose height exceeds the airport fence.

Inefficient management of the incinerator, the final disposal site for aircraft waste and the use of the airport itself.

It is recommended that a National Wildlife Hazard Management and Control Committee be established, which should include government departments such as the Civil Aviation Authority, transport, defense, agriculture and the environment, as well as representatives of the main aircraft and airport operators and flight safety officials. Educational bodies providing specialized training in bird/wildlife hazards should also be invited to participate. This Committee will put into practice the proposal for a regulation establishing the Control of Fauna in the vicinity of Airports and Aerodromes, attached to this dissertation.

## CHAPTER 5: BIBLIOGRAPHY

NATIONAL CIVIL AVIATION AGENCY (ANAC) AND NATIONAL COUNCIL OF PUBLIC PROSECUTORS. *Demands and guidelines on civil aviation*. Brazil, 1st edition, 2016.

ALLAN, J. *Bird Strikes as a hazard to aircraft*: A changing but predictable and manageable threat. United Kingdom, International Bird Strike Committee Central Science Laboratory 2000;

Annex 14 to the Convention on International Civil Aviation: *design and operation of aerodromes*, ICAO, Volume 2, 3ª ed, Montreal - Canada, 2004.

Airport Service Manual. "Bird Control and Reduction", Doc. 9137 - An/898, Part 3. ICAO, 3ª ed, Canada, 1991.

BARDIN, L. *Content Analysis*. Lisbon: Edições 70, 1995.

BITTAR, Carlos Alberto, *Curso de direito civil*. 1st ed. Rio de Janeiro: Forense, 1994. p. 561;

BLACKWELL, B. F. et. al. *Wildlife collisions with aircraft: A missing component of land-use planning for airports*, 2009. USDA National Wildlife Research Center - Staff Publications. 860. Available at: http://digitalcommons.unl.edu/icwdm_usdanwrc/860. visited on 20-09-2023;

BOGDAN, R. & BIKLEN. S. *Qualitative Research in Education*. Porto: Porto Editora. 1994.

BORG, W. R.; GALL, M. D. & GALL, J. P. *Educational research: An introduction,* New York, Longman, 2002.

CAVALCANTI, A. U. *Responsabilidade civil do transportador aéreo: tratados internacionais, leis especiais e código de protecção e defesa do consumidor*. Rio de Janeiro: Renovar, 2002.

CAVALIERI F. S., *Civil Liability Program*. 9th ed. rev. and ampl. São Paulo: Atlas, 2008. p. 3.

CAVALIERI F.S., *Civil Liability Program*. 9th ed. São Paulo Editora Atlas, 2008, p. 137.

CENIPA. *Avian Risk Management System*. SIGRA. 2017.

CLEARY, E. C. & DOLBEER, A. R., *Wildlife Hazard Management at Airports*: a manual for airport personnel. United States, 2005. Available at: http://www.birdstrike.org. Accessed February 16, 2022;

Civil Code of Mozambique, approved by Decree-Law no. 3/2006, of August 23.

CONKLING, T.J. et al. *Impacts of biomass production at civil airports on grassland Bird conservation and aviation strike risk. Ecological applications: a publication of the Ecological Society of America*, year 5, n. 28, 2018. p. 1168-1181.

DENZIN, N. K. & LINCOLN, Y. S. *Introduction: the discipline and practice of qualitative research. In*: DENZIN, N. K. & LINCOLN, Y. S. (org.) *O planejamento da pesquisa qualitativa: teorias e abordagens*. 2. ed. Porto Alegre: Artmed, 2006. p. 15-41.

DIAS, José de Aguiar, *Da responsabilidade civil*. 6. ed. Rio de Janeiro: Forense, 1987, v. 2.p. 26.

FIUZA, C. *Direito Civil: curso completo,* 15 ed. Belo Horizonte, 2011, p.331.

GAMBOA, S. S. *Pesquisa educacional: quantidade-qualidade*. São Paulo: Cortez, 1995.

GERHARDT, T. E. & SILVEIRA, D. T. *Métodos de Pesquisa. 1* ed. Porto Alegre. Editora da UFRGS. 2009. Available at:

http://www.ufrgs.br/cursopgdr/downloadsSerie/derad005.pdf. Accessed on May 12, 2021.

GHIGLIONE, R. & MATALON, B. *Inquiry: theory and practice.* Oeiras, Celta Editora. 1993.

IGEA, D., et al. *Técnicas de Investigación en Ciencias Sociales.* Madrid: Dykinson. 1995.

INTERNATIONAL BIRDSTRIKE COMMITTEE. *Recommended Practices No.1 - Standards For Aerodrome Bird/Wildlife Control.* Warsaw: IBSC. 2006.

INTERNATIONAL CIVIL AVIATION ORGANIZATION, *Airport Services Manual, Part 3 - Wildlife Control and Reduction* (Doc 9137), ICAO, Fourth Edition, Montréal, Québec, Canada, 2012.

INTERNATIONAL CIVIL AVIATION ORGANIZATION, *Manual on the ICAO Bird Strike Information System* (IBIS) (Doc 9332), ICAO, Third Edition, Montréal, Québec, Canada, 1989.

INTERNATIONAL CIVIL AVIATION ORGANIZATION, "*Manual de Servicios de Aeropuertos - Reducción Del Peligro Que Representan Las Aves,* Doc 9137 NA/898 Parte 3, 3ª ed, 1991;

MATTAR, F. N. *Pesquisa de marketing.* 3.ed. São Paulo: Atlas, 2001.

MATTOS, L. M. *Avian danger in the state of Rio de Janeiro: application of the homogeneity analysis method.* Niterói, 2014. p.118.

MELO, C. A. B., *Curso de Direito Administrativo,* 4th ed., Manheiros, São Paulo, 1993.

MENDONÇA, F. A. C. *Handbook of Avian Danger,* 2005.

MENDONÇA, F. A. C. *Avian Hazard Management at Airports, SIPAER Connection Review,* 2009 edition.

MOREIRA, D. C., *Planeamento e Estratégias da Investigação Social.* Lisbon: ISCSP, 1994.

MUCHANGOS, A. *Cidade de Maputo-Aspectos Geográficos.* Maputo, 1994. pp. 11-94.

NORONHA, F. *Direito das obrigações.* 3.ed. rev. and updated. São Paulo: Saraiva, 2010.

QUIVY, R. & CAMPENHEOUDT, L. *Manual de Investigação em Ciências Sociais.* Lisboa. Gradiva, 2003.

RAMOS. *Avian danger, prevention and combat.* Course Conclusion Paper, Universidade Tuiuti Paraná, 2009.

RAUEN, F. J., *Roteiros de Investigação Científica.* Tubarão: Editora da Unisul, 2002;

Republic of Mozambique, Boletim da República: Civil Aviation Law of Mozambique. In: Boletim da República, 5/2016 of June 14, 2016.

Republic of Mozambique, Boletim da República: Regulamento de Aviação Civil de Moçambique, which publishes MOZ-CAR PART 12 on Investigation of Accidents and Incidents with Civil Aircraft, August 30, 2018.

Republic of Mozambique, Boletim da República: Ministerial Diploma, which publishes the Civil Aviation Regulation MOZ-CAR PART 139 on Construction, Licensing and Certification of Aerodromes. In: Boletim da República, 51/2013 of May 29, 2013.

REPUBLIC OF MOZAMBIQUE, *Aeronautical Information Publication.* 3rd ed. FQMA AD 2-7, 2021.

SILVA, P., *Vocabulário Jurídico Conciso.* 1 ed. Rio de Janeiro. Forense, 2010. p. 642.

STOCO, Rui. *Treatise on civil liability: doctrine and jurisprudence.* 7 ed. São Paulo, Editora Revista dos Tribunais, 2007. p.114-157.

THORPE, J.*100 years of fatalities and destroyed civil aircraft due to bird strikes.* In International Bird Strike Committee Meeting, 2012.

TUCKMAN, B. *Manual de Investigação em Educação - Como Conceber e Realizar o Processo de Investigação em Educação*. 2ªedição. Lisbon: Calouste Gulbenkian Foundation. 2002.

VILLAREAL, L. M. A. *Programa Nacional de Limitación de Fauna en Aeropuertos*. Republic of Colombia - Special Administrative Unit of Civil Aeronautics, Colombia, 2008.

**WEBSITES CONSULTED:**
https://pt.linkedin.com/pulse/risco-de-fauna-na-avia%C3%A7%C3%A3o-birdstrike-eduardo-ortiz-pereira, visited on 19-09-2023
https://bashny.net/t/es/46036 visited on 20-09-2023
http://www.cenipa.aer.mil.br/cenipa/index.php/component/content/article/artigos-cenipa/122-matriz-de-risco-da-fauna visited on 22-09-2023
http://ultimosgundo.ig.com.br/brasil/2015-07-21/aviao-colide-com-ave-em plen-voo-e-assusta-passageiro.html/ visited on 22-09-2023

**Appendix 1**

## QUESTIONNAIRE

This questionnaire is intended for research into *"The Influence of Environmental Conditions on the Occurrence of Aviation Accidents Caused by Birds and Measures to Prevent Avian Danger: The Case of Maputo International Airport"* for the Master's dissertation in Environmental Risk Management at the University of Pedagogy - FCTA by researcher Célia José Balate Langa. The information collected in this questionnaire is reserved, anonymous and confidential, and will be used for statistical treatment.

Thank you for taking the time to answer this survey.

*Thank you very much for your cooperation.*

**I.      Personal Data**

The next group of questions concerns the characteristics of the respondents. Choose one of the following options:

| 1.Gender:<br>  1.  ( ) Female<br>  2.  ( ) Male | 2 Nationality:<br>  1.  ( ) Mozambican<br>  2.  (      )      Other      Which? |
|---|---|
| 3. Age:<br>  1.  ( ) Between 15 and 24 years old;<br>  2.  ( ) Between 25 and 34 years old;<br>  3.  ( ) Between 35 and 44 years old;<br>  4.  ( ) Between 45 and 54 years old;<br>  5.  ( ) Between 55 and over | 4. educational qualifications<br>  1.  ( ) No education;<br>  2.  ( ) Elementary school;<br>  3.  ( ) Secondary education;<br>  4.  ( ) Degree;<br>  5.  ( ) Master's degree;<br>  6.  ( ) Doctorate; |
| 5.Function:<br>  1.  ( ) Pilot<br>  2.  ( ) Air Traffic Controller<br>  3.  ( ) Airport Manager<br>  4.  ( ) Airport Operations<br>  5.  ( ) Airport and Environmental Safety Manager | 6. seniority in the profession:<br>  1.  ( ) Up to 1 year<br>  2.  ( ) 2 to 5 Years<br>  3.  ( ) From 6 to 10 Years<br>  4.  ( ) More than 10 years |
| 7. What area do you work in?<br>  1.  ( ) External to the airport building<br>  2.  ( ) Internal to the airport building | |

| 3.  ( ) Both | |
|---|---|

    **II.     Part**

**2.1 Have you seen any animals in the Mavalane airport area in the last 4 years?**

    1.  ( ) Yes                  **2.**  ( ) No

**2.2 Have you attended a lecture or training session at the airport on the risk of collision between aircraft and wildlife in the last 5 years?**

    1.  ( ) yes                 3.  ( ) I don't remember

    2.  ( ) no

**2.3 Have you received guidance on procedures to prevent animals from approaching the airport in the last 5 years?**

    1.  ( ) yes                 **3.**  ( ) I don't remember

    2.  ( ) no

**2.4 Have you attended a lecture at the airport on how waste should be dealt with in the last 5 years?**

    1.  ( ) yes                 **3.**  ( ) I don't remember

    2.  ( ) no

**2.5 How often have you seen animals on the airfield?**

    1.  ( ) Never            3.  ( ) Often

    2.  ( ) Rarely          4.  ( ) Always

**2.6 After spotting the animal in the last 4 years, what did you do?**

    1.  ( ) Nothing, just saw        6.  ( ) killed the animal.

    2.  (  ) warned an airport employee        7.  ( ) Activated environmental agencies

    3.  ( ) took home.        8.  ( ) scared the animal away

    4.  ( ) fed the animal.        9.  ( ) Other, please describe:

    5.  (  ) called the fire department

**2.7 Have you ever experienced an incident or accident involving birds hitting aircraft?**

    1.    ( ) Yes

    2.    ( ) No        3.    ( ) I don't remember

**2.8 If yes, what were the effects of the collision?**

    1.  ( ) None        3.  ( ) Emergency landing

    2.  ( ) Aborted take-off        4.  ( ) Other

### III. Risk assessment

**3.1 What are the main environmental problems faced at this airport?**

    1.  ( ) Effluent treatment        3.  ( ) Noise

    2.  ( ) Existence of fauna at        4.  ( ) Solid Waste

       the airport        5.  ( ) Other

______________________________________________________________

______________________________________________________________

______________________________________________________________

________________________________________

**3.2 Do the environmental conditions at and around the airport influence the occurrence of aviation accidents caused by birds inside the airport?**

    1.    ( ) Yes        3.    ( ) I don't know

    2.    ( ) No

**3.3 What are the main factors behind these collisions?**

    1.    ( ) Disordered land use

    2.    ( ) The existence of rubbish dumps, markets, stalls, etc. around the airport

    3.    ( ) The existence of a snack bar inside the airport

    4.    ( ) The existence of cultivated or planted areas around the airport

    5.    ( ) Deficiency of public policies for the removal of activities close to the airport

    6.    ( ) Other factors.

**3.4 What is the frequency of collisions between birds and aircraft at the airport?**

1.    ( ) Once a week

2.    ( ) More than twice a week

3.    ( ) Five times a month

4.    ( ) More than five times a month

5.    ( ) More than 10 times a month

**3.5 What measures are taken to control avian danger at the airport?**

1. (     )Changing flight schedules

2. ( ) Changing the habits and habitats of birds

3. ( )Sound deterrents

4. ( ) Visual deterrents

5. ( )Chemical repellents

**3.6. Are the Fauna Risk management measures adopted at the aerodrome efficient?**

1. ( ) Not efficient

2. ( ) Low efficiency

3. ( ) Medium efficiency

4. ( ) High efficiency

5. ( ) No measures known

3.7 Do you have any suggestions for improving wildlife conservation at the airport and reducing the risk of aircraft colliding with wildlife?

_______________________________________________________________

_______________________________________________________________

_______________________________________________________________

_______________________________________________________________

_______________________________________________________________

_______________________________________________


3.8 What actions the airport has taken with regard to wildlife, considering its responsibilities.

_______________________________________________________________

_______________________________________________________________

_______________________________________________________________

_______________________________________________________________

_______________________________________________________________

_______________________________________________

**QUESTIONNAIRE**

This questionnaire is intended for research into *"The Influence of Environmental Conditions on the Occurrence of Aviation Accidents Caused by Birds and Measures to Prevent Avian Danger: The Case of Maputo International Airport"* for the Master's dissertation in Environmental Risk Management at the University of Pedagogy - FCTA by researcher Célia José Balate Langa. The information collected in this questionnaire is reserved, anonymous and confidential, and will be used for statistical treatment.

Thank you for taking the time to answer this survey.

**Airport Profile**

1. Airport history (year of creation, among other aspects).
2. How many passengers and flights (domestic/international) per year?
3. How much cargo in tons/year?
4. What are the main environmental problems facing this airport?
5. What do you think is the biggest difference between this airport and others in terms of dealing with environmental problems?

**Environmental Management**

1. How is the airport's environmental team formed? What is their position/function and training?
2. What environmental action programs have been developed at the airport?
3. How long does each program last?
4. What is the purpose and audience of each program?
5. What actions have been carried out for each program?

6.  Is there any communication with other airports about the actions taken? If so, how does this happen?

7.  Are any programs carried out in external partnership? If yes, which one?

8.  What are the results of the programs listed?

9.  Is there a Pilot Project in the environmental area being implemented at this airport?

10. What are the airport's future environmental commitments?

11. How do you assess the influence of environmental conditions on the occurrence of aviation accidents caused by birds at the airport?

12. What are the main factors behind these collisions?

13. How often do birds collide with aircraft at the airport?

14. What measures are taken to control avian danger at the airport?

**Environmental Education**

1.  Are there any Environmental Education Programs at the airport? If so, which ones?

2.  How long does each program last?

3.  What is the purpose and audience of each program?

4.  What actions have been carried out for each program?

5.  Is there any external partnership in carrying out these actions?

6.  How do you communicate with the population around the airport? Do you carry out any environmental education activities with this public?

7.  Is there communication between the actions carried out at this airport and other airports?

8.  What are the results of the actions carried out in the area of environmental education?

Annex 1: Fauna risk assessment matrix, using Methodology 1.

| Species | Population | Pasta | Average number of individuals sighted in flocks | Presence of surveys in which the species was identified. | Period of the day (% of times sighted during the period of greatest activity at the aerodrome) | Location (% of times located in the highest risk areas) | Flight / activity ( % of times sighted in flight or intense activity) | Register (History of wildlife collision reports) | Sum |
|---|---|---|---|---|---|---|---|---|---|
| | A | B | C | D | E | F | G | H | A+B+..H. |
| Herons | 3 | 2 | 3 | 3 | 3 | 3 | 3 | 3 | 23 |
| Caspian tern/ great tern | 3 | 2 | 2 | 3 | 3 | 3 | 3 | 3 | 22 |
| Ardea Melanocephala/black-headed heron | 3 | 2 | 3 | 3 | 2 | 2 | 3 | 3 | 21 |
| Burhinis capensis / Cape Caraway | 2 | 3 | 2 | 3 | 2 | 3 | 3 | 3 | 21 |
| Stork | 2 | 3 | 2 | 3 | 3 | 3 | 2 | 3 | 21 |
| Accipiter minullus/small hawk | 2 | 1 | 2 | 3 | 2 | 2 | 2 | 2 | 16 |
| Swallow | 3 | 1 | 3 | 2 | 2 | 2 | 1 | 1 | 15 |

| | | | | | | | | | |
|---|---|---|---|---|---|---|---|---|---|
| Harrier Circus/ Falcons | 2 | 2 | 1 | 1 | 1 | 2 | 2 | 2 | 13 |
| Charadrius tricoloris/ Three-tailed plover | 1 | 2 | 2 | 1 | 1 | 1 | 1 | 2 | 11 |
| Thick-billed lark | 2 | 2 | 1 | 1 | 2 | 1 | 1 | 1 | 11 |
| Vidua regia/ queen whydah | 2 | 1 | 2 | 2 | 1 | 1 | 1 | 1 | 11 |
| Owl | 1 | 1 | 2 | 1 | 1 | 1 | 1 | 2 | 10 |
| Merops persicus/ Persian bee-eater | 2 | 1 | 2 | 1 | 1 | 1 | 1 | 1 | 10 |
| Polyborides typus/ | 1 | 1 | 2 | 1 | 1 | 1 | 1 | 1 | 09 |
| Terek sandpiper | 1 | 1 | 2 | 1 | 1 | 1 | 1 | 1 | 09 |

**Source:** http://www.cenipa.aer.mil.br/cenipa/index.php/component/content/article/artigos-cenipa/122-matriz-de-risco-da-fauna cited by Mattos (2014).

Annex 2: Problem species at Maputo International Airport (AIM).

| Herons / Ardea Melanocephala/ Black-headed Heron | Caspian tern/ Great Tern |
| --- | --- |
|  | |
| The black-headed heron is a large bird with a height of 85 cm and a wingspan of 150 cm, almost as large as the great egret, which it resembles in appearance, although it is generally darker. It usually feeds in shallow water, spearing fish or frogs with its long, | |

pointed beak, the species will also hunt well away from the water, catching large insects, small mammals and birds.

The black-headed egret is found in northern West Africa, sub-Saharan Africa and Madagascar.

Read more at: https://www.passaro.org/garca-de-cabeca-preta-caracteristicas-reproducao-e-alimentacao/#Caracteristicas_da_garca-de-cabeca-preta.

| Burhinis capensis / Cape Caraway  | Stork  | Accipiter minullus/ Sparrowhawk  |
|---|---|---|

| | | |
|---|---|---|
| The cape caracara, which can reach up to 45.5 cm in height, has long legs and brown and white speckled plumage that provides camouflage, making it difficult to spot the bird in the grasslands and savannahs where it roams<br>The Cape scarab hunts exclusively on the ground, feeding on insects, small mammals and lizards up to mice. This species of bird is nocturnal and crouches on the ground during the day, making it difficult to locate. It hunts exclusively on the ground, feeding on insects, small mammals and lizards.<br>The cape caraway is native to the grasslands and savannahs of sub-Saharan Africa, and can also be found from Senegal, Mali and Mauritania in the west to Ethiopia Kenya, Tanzania and South Africa in the east and south.<br><br>Source: https://www.passaro.org/alcaravao-do cabo/#Caracteristicas_do_alcaravao_do_Cabo | The stork (Ciconia ciconia) is a bird belonging to the Ciconiidae family, also called the common stork. It is characterized by its large size and has black and white plumage on its wings, and gets a red color on its legs and beak in adults.<br>The stork is a large bird, which can reach a length of 100 to 115 cm and a height of 100 to 125 cm. It weighs between 2.3 and 4.5 kg. However, like all storks, its legs and neck are long, as is its beak, which is long, straight and pointed.<br>This bird feeds on insects, fish, amphibians, reptiles, small mammals and small birds that are caught on the ground or in shallow water.<br>I'm sure you've wondered where the stork lives. This bird has a wide distribution in Europe, North Africa and some parts of | The species of the *Accipiter* genus have very varied plumages. A common feature is their slender build and very long tail, which can reach half the total length of the body. The legs are relatively tall. Like all accipitrids, they have a curved beak, suitable for feeding on small birds and reptiles. They generally live in forested areas and have a worldwide distribution.<br>It can be found in the following countries: Angola, Botswana, Burundi, Democratic Republic of Congo, Eritrea, Ethiopia, Kenya, Lesotho, Malawi, Mali, Mozambique, Namibia, Rwanda, Somalia, South Africa, Sudan, Eswatini, Tanzania, Uganda, Zambia and Zimbabwe. |

| | | |
|---|---|---|
| | western Asia. Migratory routes expand the range of this species to many parts of Africa and India.<br><br>Read more at: https://www.passaro.org/cegonha/#Caract eristicas_da_cegonha | |

**Annex 3:**

**Proposal for a Decree Approving the Regulation Establishing the Control of Fauna around Airports and Aerodromes**

BACKGROUND

Volume I of Annex 14 to the Convention on International Civil Aviation, commonly referred to as the Chicago Convention, requires that the risk of collision with wildlife on or near an aerodrome be assessed through, among other things, the establishment of national procedures and a continuous assessment of wildlife and its hazards by competent personnel.

The above-mentioned Annex states that it is necessary to create a national committee for verifying the risk associated with wildlife in the vicinity of an aerodrome, as they have proven to be specialized or technical forums for obtaining and exchanging information on research and development in wildlife control at the airport.

In this sense, the committee should include all the stakeholders associated with or interested in the problem, since they are multidisciplinary and generally act as a source of information for members of the aviation community.

A national committee should include government sectors such as transport, defense, agriculture, the environment and educational institutions, as well as representatives of the main aircraft and airport operators, flight safety officials, pilot associations and airframe and engine manufacturers.

Given that collisions with birds/wildlife represent a persistent problem, studies should be carried out to assess the danger posed by wildlife on a case-by-case basis.

National guidelines or regulations should be developed as a basis for the committee, as well as for the direction of airport authorities, aircraft operators and other institutions. These guidelines can also provide the basis and mandate for the development of special bird/wildlife research and control programs.

Considering that the increasing proliferation of degraded areas and the lack of basic sanitation in areas close to airports are conducive to the incidence and permanence of birds in these areas, this could culminate in an increase in aircraft collisions with birds.

Law No. 5/2016 of June 14, Mozambique's Civil Aviation Law, in its article 33, establishes that the use of properties and facilities neighboring aerodromes and heliports, as well as air navigation support facilities, covered by aeronautical easements, are subject to special restrictions aimed at ensuring the safety of air operations, and the restrictions referred to in the previous paragraph are related to:

a) use of property for the construction of buildings, agricultural crops or other purposes;

b) the presence of animals, vehicles, illuminated signs and other objects, whether temporary or permanent;

c) anything that may hinder aircraft maneuvers, cause interference with radio navigation aid signals, jeopardize the visibility of visual aids or otherwise compromise the safe operation of aircraft.

In this way, the above-mentioned law establishes that the Civil Aviation Regulatory Authority of Mozambique, in coordination with local and municipal authorities, must establish a general plan for aeronautical easements for airports, aerodromes and heliports, referring to specific regulations for its operationalization, this proposal for a decree is submitted which aims to establish the Control of Fauna in the vicinity of Airports and Aerodromes.

# REPUBLIC OF MOZAMBIQUE

## COUNCIL OF MINISTERS

---

Decree no. / of 2023

In view of the need to establish the rules aimed at reducing the risk of accidents and aeronautical incidents arising from the collision of aircraft with specimens of fauna in the vicinity of aerodromes, in the exercise of the powers conferred on it by <u>article 93 of Law 5/2016 of June 14</u>, the Civil Aviation Law, the Council of Ministers hereby decrees:

Article 1 The Regulations for the Control of Fauna in the vicinity of Airports and Aerodromes are hereby approved.

Article 2 This Decree shall enter into force on the date of its publication.

Approved by the Council of Ministers, at ________ de _____ de 2023.

Publish.

The Prime Minister, *Adriano Afonso Maleiane.*

# REGULATION ON THE CONTROL OF FAUNA IN THE VICINITY OF AIRPORTS AND AERODROMES

## Article 1
(Object)

This Regulation establishes rules aimed at reducing the risk of accidents and aeronautical incidents resulting from aircraft colliding with specimens of fauna in the vicinity of aerodromes.

## Article 2
(Definitions)

For the purposes of these Regulations, the following are considered:

1. slaughter: the killing of animals at any stage of their life cycle, caused and controlled by man;

2. aerodrome: any area intended for the landing, take-off and movement of aircraft;

3. military aerodrome: an aerodrome intended for use by military aircraft;

4. airport: any public aerodrome with facilities and installations to support aircraft and the embarkation and disembarkation of people and cargo;

5. Airport Safety Area - ASA: a circular area of the territory of one or more municipalities, defined from the geometric center of the largest runway of the airport, aerodrome or military aerodrome, with a radius of 20 km (twenty kilometers), whose use and occupation are subject to special restrictions due to the nature of the attraction of fauna;

6. fauna-attracting activity: solid waste dumps and any other activity that serves as a focus or contributes to the significant attraction of fauna within the ASA, compromising the operational safety of aviation;

7. activities with the potential to attract fauna: landfills and any other activities which, using the appropriate operating and management techniques, do not constitute a focus of attraction for fauna within the ASA, nor compromise the operational safety of aviation;

8.      environmental authority: governmental, provincial or municipal body or entity that is part of the National Environmental System and is responsible for granting environmental licenses;

9.      civil aviation authority: regulatory body for civil aviation (IACM);

10.     municipal authority: the competent body or entity of the municipal administration;

11.     capture: the act or effect of detaining, containing by mechanical means or preventing the movement of an animal, followed by its collection or release;

12.     problem species: a species of fauna, native or exotic, that interferes with aviation operational safety;

13.     synanthropic species: an animal species adapted to living alongside humans, despite their wishes, and which differs from domestic animals bred for the purposes of companionship, food production or transportation;

14.     fauna management: the application of ecological knowledge to the populations of species of fauna and flora, seeking a balance between the needs of these populations and the needs of people;

15.     aerodrome operator: the body, entity or company responsible for aerodrome management;

16.     suitability parameters: measures determined by the competent authority with the aim of managing and reducing the risk of aeronautical accidents and incidents resulting from aircraft colliding with specimens of fauna at aerodromes;

17.     Aerodrome Fauna Management Plan - PGFA: a technical document that specifies in detail the necessary interventions in the natural or man-made environment of an aerodrome or directly in the populations of native or exotic fauna species, with the aim of reducing the risk of collisions with aircraft;

18.     National Fauna Risk Management Program - PNGRF: a normative document that establishes objectives and targets with the aim of improving operational safety in the country by proactively managing the risk of aircraft colliding with native or exotic species of fauna;

19.  special restrictions: any of the following limitations imposed by the competent aviation authority on the use of property, public or private, located within the ASA:

a) a ban on the implementation of activities that attract fauna specimens;
b) the immediate or gradual cessation of any activity that attracts specimens of fauna, and the person responsible for the activity must comply strictly with the provisions of the environmental legislation in force, including recovery of the degraded area;

c) adaptation of activities with the potential to attract fauna specimens to the parameters defined by the competent authority, accompanied or not by their suspension;

d) the implementation and operation of activities with the potential to attract fauna specimens, subject to authorization and suitability parameters, both defined by the competent authority;

20. operational safety: a state in which the risk of injury to persons or damage to property is reduced and maintained at or below an acceptable level through a continuous process of hazard identification and risk management; and

21. translocation: capture of living organisms in a given area for subsequent release in another previously determined area, according to the species' geographical distribution.

Article 3
(Establishment of the Aerodrome Safety Area)

In order to manage and reduce the risk of aeronautical accidents and incidents resulting from aircraft colliding with wildlife at aerodromes, the Airport Safety Area (ASA) is established, where land use is restricted and subject to compliance with specific regulatory requirements for aviation operational safety and the environment.

(1) The perimeter of the aerodrome's Airport Safety Area (ASA) will be defined from the geometric center of the largest runway of the aerodrome or mixed aerodrome and will comprise a radius of 20 km (twenty kilometers).

(2) The National Fauna Risk Management Program - PNGRF, developed and supervised by the civil aviation, military and environmental authorities,

120

will cover objectives and goals common to aerodromes and their respective ASAs.

## Article 4
### (Airport Security Area)

1. "Airport Safety Area - ASA" means the areas covered by a certain radius from the "geometric center of the aerodrome", according to its type of operation, divided into 2 (two) categories:
   a) 20 km radius for airports operating under instrument flight rules (IFR); and
   b) 13 km radius for other aerodromes operating under flight rules.

2. In the event of a change of aerodrome category, the radius of the ASA must be adapted to the new category.

## Article 5
### (Prohibitions)

Within the ASA, it will not be allowed to set up activities of a dangerous nature, understood as "bird attraction sites", such as slaughterhouses, tanneries, landfills, open-air and uncontrolled garbage dumps, planting of agricultural crops that attract birds, as well as any other activities that could pose similar risks to air navigation.

## Article 4
### (Special restrictions)

1. The special restrictions contained in the PNGRF must be observed:
   a) by the municipal authority, in the planning and control of the use and occupation of urban land, which is responsible for implementing and supervising the PNGRF;
   b) by the environmental authority, in the environmental licensing process and during inspection and control activities; and
   c) by the aerodrome operator, at the airport or aerodrome administration.

(2) Rural properties incorporated into the ASA are also subject to the special restrictions laid down in the PNGRF and to supervision by the municipal authority.

(3) The municipal planning instruments governing the subdivision, use and occupation of land shall comply with the provisions of this Law and the special restrictions laid down in the PNGRF.

Article 5
(Information obligation)

Public, provincial or municipal institutions, the aerodrome operator and the owner of real estate or developments located in the ASA are obliged to provide the information requested by the civil aviation authority or the military aeronautical authority.

Article 6
(Wildlife management plan)

1. Fauna management on and around aerodromes will be authorized by the environmental authority upon approval of the Aerodrome Fauna Management Plan (PMFA) and may involve:
   I - environmental management;
   II - management of animals or parts thereof;
   III - transportation and disposal of the collected zoological material;
   IV - capture and translocation;
   V - collection and destruction of eggs and nests; and
   VI - animal slaughter.

2. The FMP must evaluate ways of controlling and reducing the potential danger of aircraft collisions with fauna specimens, supported by data obtained using scientific methods and taking into account aspects of the population dynamics of the problem species(s).

3. The slaughter of animals will only be permitted:

   a) after proof that the use of indirect and direct management of the problem species(s) or the environment has not produced significant results in reducing the danger of aircraft collisions with specimens of fauna at the aerodrome;

   b) after proving that the environmental impact or the economic cost of transferring cynanthropic species or non-endangered problem species(s) does not justify the translocation.

3. The slaughtered animals, nests and other zoological materials collected may be sent to collections in scientific institutions or discarded.

4. Zoo material must be disposed of by burial, landfill, incineration or any other appropriate means possible in the municipality where the airfield in question is located.

5. Authorization to manage wild fauna does not exempt bearers from complying with the relevant law.

## Article 7
### (Contraventions)

It is a contravention of the provisions of this Law:

1. implement or operate an activity with the potential to attract fauna specimens in the ASA without submitting it to the approval of the municipal authority, the environmental authority and the Civil Aviation Authority of Mozambique;

2. Encouraging, developing or allowing the development of activities with the potential to attract specimens of fauna considered prohibited within the ASA;

3. Failure to comply with the deadline set for the cessation of activities with the potential to attract fauna specimens;

4. Failure to adapt activities with the potential to attract fauna specimens to the parameters defined in the special restrictions; and

5. Failure to comply with the order to stop attracting fauna specimens.

## Article 8
### (Sanctions)

The following administrative sanctions are applicable for the infractions provided for in Article 7 of this Law:

    a) warning notice;
    b) simple fine;
    c) daily fine;
    d) suspension of activity;
    e) interdiction of an area or establishment; and
    f) construction embargo.

(1) Administrative sanctions shall be suspended as soon as the reasons for imposing them are remedied.

2. The sanctions provided for in points b) and d) of this article may be applied cumulatively.

3. Fines will be imposed according to the seriousness of the offense, in accordance with MOZCAR 11 on misdemeanors.

## Article 9
### (Aggravating circumstances)

Circumstances that aggravate the sanctions provided for in this Law are:

1. Recidivism;
2. Evidence that the offender, by engaging in any of the actions provided for in Article 7 of this Law, has collaborated in the occurrence of an aeronautical accident or incident resulting from the collision of an aircraft with specimens of fauna in the vicinity of an aerodrome.

Article 10
(Competent authority)

The Civil Aviation Authority is responsible for applying the administrative sanctions provided for in this law, with the opinion of the National Committee for Wildlife Hazard Reduction.

Article 11
(Destination of fines)

The amount earned from the collection of fines should be used to cover the costs of the National Committee for the Reduction of Fauna Danger (CNRPA) in activities that contribute to reducing the risk of aeronautical accidents and incidents resulting from the collision of aircraft with specimens of fauna.

Article 12
(Transitional provisions)

Existing activities of a dangerous nature within the ASA must adapt their operations in order to minimize their attractive and/or risky effects, in accordance with regulatory safety and/or environmental requirements, within ......... days of the publication of this Resolution.

Printed by Books on Demand GmbH, Norderstedt / Germany